MASTHEAD

EDITORS
Rose Alexandre-Leach
Sierra Dickey
Jenna Gersie
Anna Mullen

PUBLISHER | Dede Cummings

COVER ART | Kelsey Swintek
Film, 2015

ILLUSTRATOR | Sam Mass
Watercolor on paper, 2016

EDITORIAL SUPPORT
Ron Anahaw
Emy Blohm
Ferne Johansson
Kaitlyn Plukas
John Tiholiz

FOR QUESTIONS AND COMMENTS
editor@hoppermag.org

hoppermag.org
facebook.com/hopperlitmag
Twitter & Instagram: @hopper_mag

T0206791

LETTER FROM THE EDITORS

Silence and stillness
Vacuoles plumped to fill
Glucose readied

When *The Hopper* was germinating this October, we imagined a literary magazine of and about New England. Inspired by the hard-scrabble but warmhearted history of so many Vermont villages and the new cultures heralded as coming out of these hills, we pictured the magazine as a pure product of one region—something as vertically deep as Dillard and as socially networked as Mosher or Berry.

Our name and icon were derived from a Leland Kinsey poem. The name *The Hopper* led us to compare cider making to the literary arts—farming is never far off from other creative processes. Experiences, objects, and tastes are harvested from here and there, brought together, milled in a frenzy, and processed down to bare. Work around harvests, especially fruit harvests during banner years, is time sensitive, and the workers are often transitory. The labor around this magazine has been the same: people have entered and exited, sharing different parts of the load at different times.

This first annual print issue brought with it a litany of surprises. First off, by no particular effort (or lack of) on our part, the authors and artists we chose to publish are mostly non-native to New England, and only four of the twenty-six are resident Vermonters. What we imagined as a hyper-focused (geo-fenced, you could say) magazine about place has become multi-focused. We would have been happy with just New England, but tasting so many different vernaculars of space has been incredible. Clearly, good writing disrupts the ideas one has about one's magazine, and about oneself.

That said, we hope this issue solidifies for the concerned reader that nature writing is alive and well and only getting more nuanced and applicable in our Internet age. And, if none of the writing herein demonstrates that, we'll let slip a secret that should do it. What you thought were apples on this cover are actually green oranges, seen and photographed outside a mosque in Seville, Spain.

The Editors

The HOPPER

CONTENTS

HARVESTING

We walked the rows of the orchard, picking Honeycrisps, Cortlands, and Ida Reds. Twisting MacIntosh and Macouns from their stems, we piled the fruits in the white half-peck bag, sturdy despite its paper straps. We collected apple vocabulary like we collected the apples: "robust with a zesty, pineapple citrus," or "crisp, juicy, and sweet with a distinct strawberry flavor." We collected apples like we collect words, like we collect poems, like we collect stories.

STEPHEN SIPERSTEIN

First Chanterelle by an Old Hunting Road

I've said a prayer
sharpened a knife
prepared myself to feel
like a fungus.
Then this golden
apparition happens
so quick
I think this,
this will be easy,
and slice the stipe
just above dirt.
But day empties
into dusk
and I've wandered
already so deep
into woods riddled
by bullet holes
and bleached bones
that lurk beneath ferns,
now far from that
initial spot
having forgot
the slope, the soil
the light, the slow
hunger of autumn
growing inside me:
a craving
for thick brown bread
dipped in melted butter
for the smell of apricot
for a feeling
like a startled
animal that jumps
from the underbrush
of the heart, crying:
"Go on, go on."

Wishing Life | MEGHAN RIGALI
Pen and ink, pencil, gel medium, acrylic, and silk of the milkweed pod, 3 x 4 ft., 2005

Burn It, Bury It, Send It Downriver | NONFICTION

BURN IT

The side of our property not bordered by farm fields and trees is intertwined with the cat ladies' place. Two older sisters babysit their grandkids before and after school. Mid-morning, one sister in a floral muumuu stands at the edge of the porch and throws cat food into the yard for the bazillion cats that live around their house, seeking shelter in the garage they leave open all year round. Garbage sits in bags at the bottom of their deck steps. My old beagle would sniff her way over there and take advantage as the scavenging opossums did the night before. Bones and Tastykake wrappers and stale bread were what she usually brought back to our yard and guarded like pirates' treasure. Once she found half a moldy pound cake to munch in the shade of our maple tree.

Two to three times a week, one of the dads dropping off his kids will place the garbage bags in a burn pit thirty feet from our house. Most yards around here contain a large discarded oil drum known as a burn barrel, or a pit that's sunk into the ground. My neighbors' burn pit is more of a clumsy tower. Cement blocks are stacked four feet high. They break and pop and send the tower tumbling down, just enough to make it both satisfying and frightening. But this rarely happens. It takes a fire to create such heat. There's rarely a fire. Usually the bag catches a flame, the man drives away to his day job, and the trash smolders for the next twelve hours.

The property line we share with the cat ladies is not straight. It makes turns to form odd angles, puzzle piecing our lands together. Both houses are nestled in a little hollow, and the wind funnels through it from them to us. On burning days, smoke envelops our house, blowing across the windows for our bedrooms and the living room where we spend much of our time. We can't turn off the air conditioner and enjoy open windows. We can't play outside.

Burning garbage is not illegal around here. I checked. It's not illegal in most of the county, unless you live in a town where garbage removal is a municipal service. With a population of 39,702, less than half the residents of my county live in a town.

I feel good about myself, a sense of smugness, for recycling even though it's inconvenient, for composting even though it's smelly, for paying Hometown Disposal to pick up what's left, for not being just like my neighbors.

Hometown Disposal is located in the county across the river, home to the largest man-made mountain in America. A mountain made from the detritus of mining anthracite coal. You come around a bend in Route 61 and there it is, blocking your view through the valley. It's still growing, something the locals say with pride and pity. Yellow earth movers scuttle up the roadways, small like little toys up so high, twisting around the plantless mountain.

Continue on 61 to Centralia, a town evacuated by the federal government in 1964. It sits atop a mined coal vein that caught fire in 1962 and still burns to this day. The biology departments of the local universities conduct research there, each claiming a plot of smoldering ground, tracking fluctuating temperatures and bacteria growth.

Route 61 was redirected after part of it collapsed into the fiery earth. Drive through in the summer, you can see patches of brown grass scorched from below. After a spring rain, the mist of evaporating water rises to rejoin the atmosphere. In winter the snow's blanket is incomplete, melted away at the hot spots. The feds tore down all but three of the houses, the ones where a few people refused to take the bailout and move to New Centralia. The grid of streets and sidewalks remains, the plants slowly reclaiming ground every summer. The shrine from an old Catholic church looks over the town from the side of the mountain. Hollywood made a horror movie about it.

The unofficial cause, the one everyone around here swears is true, involves trash burning. A guy was out back of his house, innocently firing up the pit and burning the week's garbage, when the ground opened into the mine tunnel. Once flames caught the vein, there was no return.

Bury It

I never see the Amish burning their garbage. Not that I'd assume they generate much waste, but our Newfoundland taught me otherwise. She got away from me this past winter, running across the road into an Amish cornfield. I chased after, my heart racing for fear that someone would shoot her. She looked just like a bear, romping through the broken stalks, freshly spread manure clumping in her fur.

She thought we were playing the Momma-chases-me game. Up through the field we went until she stopped at the wooded hilltop. Securing her leash, I realized she was stiff and staring. Two cows lay just yards away. Their stomachs bloated with the gas buildup of a decomposing intestinal tract. The jaw of one missing its skin. No smell. They were fresh.

Like any adult millennial, I asked my dad why it had happened. I'm originally from an area not unlike where I live now, and I assumed he'd know. He said the farmer was just waiting for the ground to thaw so he could bury them. "Why not butcher and eat them?" I asked.

"An old dairy cow's no good for eating." But why bury them? Isn't there some service to take them away? "Not really. It's probably what they've always done. I mean, what else are they supposed to do?"

Growing up in rural Pennsylvania, I've always understood you can mine the earth mound of a farm's tree lines or fence rows to find glass Coke bottles and tin tobacco cans from the early 1900s. Any packaging that couldn't be reused or burned ended up buried, and those luxuries didn't come to rural America until the last century. Waste management, the concept of thinking about and regulating our waste as a government service or a for-profit enterprise, is relatively new to us too. Stories of waste and its connection to public health serve as turning points in the modernization of our country. Modernization is something that belongs to cities. Around here people just take care of what needs to get done.

Rural Americans are so far behind that we've be-come fashionable again. When the environmentally conscious urbanite looks for knowledge they lost two or three generations ago, they come to us. We're still doing things the old way. Organic farming, canning, composting, all the DIY oils and balms fashionable once again are daily life for us. When you live in a place were no one thought to run electricity until FDR's New Deal, is it any wonder that we might be making our own landfills?

One day my family walked in the opposite direction of Dead Cow Hill, up through the fields behind our house. At the top is a wooded ridge running the mile-length of our road. Where the trees start, the trash starts. Old refrigerators, broken farm equipment, twisted metal pointing towards the sky promising tetanus and impalement. Disappointing but not unfamiliar.

We climbed to the high line for safety and the views, but found a thick plastic spine tucked alongside the ridgetop, extending out of sight. We assumed it was a row of hay, encased for the coming winter, a common sight in fields through the valley. But those plastic-covered hay snakes are always white and on flat land close to barns. This one was black and would be hard to reach in snow. "I don't understand," my husband said, keeping his distance. "What is that?"

"It's trash bags," I said.

They were. They still are. One gigantic line of black trash bags, piled tight. Decades worth of garbage. A giant, dirty secret that grows and grows.

Send It Downriver

All of this bothers me. Now I judge. I assume people must be totally ignorant or just environmental ass-holes. But maybe they're as desperate as I used to be.

I rented a place on the river for four years, just a few miles from where I now live. I moved there with my first husband. He wasn't really making a go of being a responsible adult, but I loved him and thought that would be enough.

We moved to be closer to the college where I worked as an administrator. Gas rose to over four dollars a gallon that summer. I made a decent salary, but not enough to support two heavy-drinking adults with dogs and substantial college loans and an hour commute. He worked a part-time seasonal job at a nearby golf course. Our new home was an old trailer. The kitchen counters were black, the walls dark wood paneling, and there was this trippy track lighting in the ceiling. Our old hippie of a landlord used the property to store junk he couldn't bear to get rid of, and we didn't demand it since we were desperate and he wasn't asking for much in the form of rent or credit checks. Plus the place sat atop a garage on the banks of the Susquehanna River, just below the confluence of the north and west branches. The water there is wide and flat. We owned kayaks. The property had its own dock.

The neighborhood was known as Shady Nook. It had one road in and out. I had never heard of it until we went to look at the place, and even then I couldn't believe it existed, hidden away but still less than a mile from the mall. When I told colleagues, they'd say, "I think I've heard of it, but where is it?" as if the Nook was a Neverland you couldn't get to unless shown. People partied every night. Fireworks rarely caught the attention of police. The constant barrage of mayflies and heavy humidity kept everything rotting and dirty, like you were camping. You just couldn't keep clean. I loved it.

Hometown Disposal picked up our garbage for a while, but then they stopped. Their bill was the only one that came in the mail, a paper bill requiring a check and envelope and stamp that had to be gathered and assembled and my god it just seemed like such a nuisance to me. I put my husband in charge of it. Made him responsible for one bill.

Spoiler alert: he didn't pay.

I told him he still had to take care of the garbage, and I didn't care how. He built quite a fine fire pit, sunk into the ground and lined with fireproof stone, and burned our trash. Just like all our neighbors. It was Jack, the guy next door, who taught my husband to construct a solid fire pit. He knew all the pits in the neighborhood, could tell just by looking around at night who was working with what. He even taught us how to smell what was being burned.

But not everything is meant to be burned. And if you couldn't burn it, you could send it downriver.

All manner of things went floating by our yard, entertainment while we sat, drunk, by the fire pit after work. We'd call out what we saw, especially exciting a few days after heavy rains when things from upriver had a chance to make their way down. The game spun from our time kayaking with a friend, who taught us to call out the trash we floated over. Golf ball! Bicycle! Refrigerator! Real estate "For Sale" sign! Golf ball! Mattress! Another golf ball. Car?

I know it sounds awful. After a while you feel like there's no use in caring anymore, in treating the river right. My husband used to drive golf balls into the water, pissing me off because I couldn't come up with a good enough argument for why he should stop. "Tell me how this golf ball is any worse for the fish than all that fracking water being dumped up north?"

The night my husband left me, he said I had matured into such a strong, independent woman that he was too intimidated to stay. Painful, but quite the reprieve.

We split, amicably, even though no one believes that's a real thing. But it was for us, and that included striking a bargain about the splitting and storing of material goods. He moved up north to work on a natural gas drill and took almost nothing with him. And we had a lot of shit. Scraping by financially meant we tended to hold onto anything and everything offered to us: freebies at golf course events, hand-me-downs from older siblings upgrading to houses, left-behinds from friends who moved out of the area. Each of our parents had downsized upon retirement, so we carried all the boxes of shit you think you want to save when you're younger: your first ball glove, third-grade classroom photo, that project you did about sea anemones in seventh grade that comes complete with a cardboard and yarn replica, high school yearbooks, every fucking book you bought for college classes, and clothes you swear you'll be skinny enough to fit into again one day, especially the skanky ones you wore to fraternity parties because those will totally be handy when you're in your thirties. They all stayed in the garage.

Working full time and in grad school part time, I had to take over his household chores, including the trash. Hometown Disposal wouldn't pick up my garbage. I still bore his last name and they had severed ties for good. I had to figure out what the fuck I was supposed to do. I took all organic waste to a compost pile that sprouted zucchini in the summer. I diligently recycled, finding empowerment in the careful sorting and storing of items until the bins in

High Tension Juice Heading for LA, Hoover Dam | WILLIAM C. CRAWFORD
Digital photography, 2015

town were open for drop-off. I put what little trash I generated into the kitchen garbage can. When the trash became level with the top of the can, I'd push it down as hard as I could, compacting it for a few more days of delay. Eventually that wouldn't work anymore. It would rise like a sponge, filling the space, demanding to be emptied.

I tried burning it at first, the relief accompanied by guilt. The second or third time I burned, an aerosol can exploded. The boom was like a gunshot, and the explosion mushroomed trash out of the pit. Bits of black debris flew twenty feet in the air and floated slowly into the yard, still burning. It was enough to bring the doctor's family, in a vacation home two plots down, out onto their deck to stare and look concerned. If I hadn't been half drunk, I might have been more scared for my safety. Instead I simply cursed my ex and decided it was his fault for putting me in that position.

The explosion didn't stop me. I couldn't stop generating trash, not in modern America. So the next weekend, in an effort to clean up and reclaim my living space from the mess we'd made of it, I pushed an ancient loveseat onto the deck, over the railing, and into the pit. I set that bitch on fire and watched as the flames caught the upholstery, shooting more than forty feet in the air, licking the limbs of the giant oak that gave me shade as I drank by the river. The heat was so intense that I moved to the edge of the property to escape it. For a moment I panicked, wondering what to do if it got worse. It didn't. Jack, conveniently also a member of the local volunteer fire company, came out with a beer and let me know that upholstery went particularly hot and fast. I was in the clear this time, but probably shouldn't do that again. "There are some things you just shouldn't burn," he said.

So I squirreled my garbage into grocery bags, taking one to work every few days to throw in a can at the edge of the parking lot. If asked, I would say I was simply cleaning out my car and happened to have a spare bag to put stuff in. When I went through my kitchen cabinets and had a box of foodstuff too old to donate, I took it to the local grocery store, thinking it wouldn't look out of place in the cans outside the main entrance, right? A grocery bag with old litter got thrown away at the pet supply chain store. And

on and on. Returning things to where they came from made sense in the game of justifications I played in my head. Yet it felt dangerous. I was petrified I'd be caught. Especially at work. There were cameras in the parking lot, and although I would remind myself that no one was getting paid enough to care about my garbage habits, I knew it would be embarrassing and hard to explain away. Adrenaline thumped every time I made that short walk, bag in hand, from my car to the can.

I filed for divorce and needed to get away, to start somewhere new. I sent résumés into the world, interviewed, and even received an offer from a college several hours away. But they had to increase their initial salary offer just to match what I was currently making, so there wasn't much financial sense to it. And then, with the opportunity at my feet, I realized I had no plan, money, or ability to move all that shit, mine and my ex's, that filled my garage. I knew it was useless, had become trash to me, but I was without access to services that could remove it. I'd stay awake at night and endure waves of anxiety when I thought about it. The weight was suffocating.

A month later, Hurricane Isaac came up the coast from Florida and rode the Susquehanna River right into our valley, hanging out, dumping for days. It was the second worst flood in history, just inches below the crest of infamous Agnes from '72 that I'd heard about my whole life. Shady Nook was evacuated. I fled to my parents' house, two hours to the south.

We were allowed back in the Nook three days later on a Sunday afternoon. I told myself it would be fine. Some clean-up work and a little elbow grease. I'd volunteered in New Orleans after Katrina, so I was prepared to handle whatever I found. I planned to arrive, get the dogs settled inside the trailer, and see what was what.

I could barely get to my place. I spent the first two hours shoveling mud from the driveway. The smelly sludge was covering the whole property two inches deep. I broke out the snow shovel and tried not to gag as I scooped up rotting fish. Keeping it off the dogs felt like a joke, but I tried anyway. Several tubs of water and old towels sat on the deck for weeks until the saturated yard could fully absorb the mud. I'd place the dogs' feet in one tub, wipe with a towel, place them in another tub, wipe again, and keep them calm with a song I sang about washing our feet, washing our feet, we're just washing our feet.

When I looked beyond the mud, I saw what was missing. The trash. The rusty portable hammock frame I brought from my childhood home, broken flower pots, barrels the landlord left behind—they were all gone. The small scrapyard from under the deck was missing, the water having wiped it clean, giving me a new start. Jack called over the fence to ask how I was making out, to see if I was missing anything. I wasn't. Things were gone, but I missed none of it. "We lost the picnic table," he said. "But it was never ours. Showed up after the last flood, so the river's taken it downstream for someone else to use."

I waited until the next morning to open the garage door. Scrappers were already parked on the street, walking from house to house inquiring about unwanted metal. I stood, staring at the shit I didn't want, no longer boxed and stacked in an orderly fashion. It spilled out, the garage a beast with its belly split, intestines everywhere. Sterilite containers, emptied of their contents, were filled with a murky liquid that had a rainbow sheen on the surface. Black goo, loosed from one of the miscellaneous cans the landlord left on a shelf, covered things with long strands like the Silly String children play with. "Got anything I can haul away for you?" a scrapper asked me, rocking from foot to foot, surveying my pile of garbage.

"I have no idea."

It was the end of summer, still hot and humid, so I needed to get everything out from under the trailer before it grew mold. I spent a day carrying everything from the garage to the turnaround of the driveway. The pile grew in small but satisfying increments. A Red Cross truck came through, handing out buckets filled with cleaning supplies and pamphlets about shocking your well and working with FEMA. The mailman came through, delivering mail to your box if you still had one, hand delivering if not. None of the neighbors went to work. Jack and several others occasionally congregated on the road, sharing information about who needed what, where everyone was,

and what needed to get done. After past floods the township placed a large container dumpster at the end of the road to collect trash. They decided this would surely be done again.

The youngest son of one neighbor came home from college to help with the cleanup. He asked me to save any scrap metal for him and a buddy to take and make some extra money. I told him my landlord left a lot of broken things behind and they were all his now. I even had a clothes dryer I'd never figured out how to hook up and now that was dead too. After he loaded up and hauled it all away, he told me I'd made him $173. "If you don't mind me asking, why did you have all that junk in your garage?" I told him I didn't even know.

The township did bring a dumpster. My landlord even came with his pickup truck and helped haul my trash pile to it. My ex came for a day with a case of beer and helped for several hours, saying "Most of this shit's mine anyway." Our divorce went through. I got my maiden name back and arranged for garbage pick-up. Then I met a guy and felt like a fully functioning adult with him. We bought a house, had kids, and got married. I'm being all responsible and shit.

Recently we had excavation work done to control some drainage issues around our house. We asked the crew leader to take his bulldozer and push back an unruly tree line that prevented our yard from having a smooth edge. He unearthed a staggering and unsurprising amount of trash: the metal frame to a screen door, a broken window, three rusted baking pans, a half-burned woven poncho like stoners wear. An assortment of degraded plastics fill in the gaps like mortar in a brick wall. It's so overwhelming that we have yet to tackle it.

Sometimes I get mad and pull out whatever's in easy reach to send it away with the trash. I grab one thing, intending just to take a small piece of the whole, but it's hard to stop. I pull and pull until my hands are full. I pull until the weight reminds me, again, that there is only so much I can do. ❧

Whooping Cranes with Polluted Sky | JENNY KENDLER
Graphite, gouache, and watercolor on paper, 16 1/4 x 12 1/4 in., 2010

Pishing

Rain is the gentlest trigger to drink:

you believed in hillocks
and not hills

and this is just creed
but not any cause

for burgeoning,
for all these bright banners for place

Here I'm teasing out the long peal

      the single *peeew*

of a goldfinch

because this is the one sound

for the blue space
in between the topped trees

The simian beards us
names us, makes us measure seasons

as plateaus of dust
when only the mountains
move us from thirst to wet

I cannot reckon
our unlikeness to hives

but except for the draping hems of cranes
that hold us to earth

the rudder beneath us is flesh,

apparent.

Wonderberry Jam | NONFICTION

MOM AND I are at a specialty seed store in the city: I am expanding my gardening repertoire, venturing into fertilizer and heirloom seeds. I see a carton of moss-growing gel, photos of melons I didn't know could grow in Alberta, seeds for familiar flowers in exotic colors, and then—sunberries. Mom finds them on the shelf, holds them out to me, tells me that in all her years she has never actually seen a seed packet of them for sale before. Of course I buy them.

They are labeled sunberries on the front, but the phrase "also known as wonderberries" on the back is a thrilling confirmation of my Internet research years before. Grandma never called them by either name; to her, they were always known as *schwarzbeeren*.

There wasn't much we could afford to gorge ourselves on when I was a kid, but we could eat as many berries from the garden as we wanted. Smatterings of fine gray garden soil speckled the eggplant purple of the berries, kicked up from water droplets hitting the earth around the plants, but that didn't matter to me. I would hold the plant with one hand, close my other hand over a bunch of berries, then tug them from their tiny star-shaped sepals. If it was hot and the skins were stretched with warm juices inside, one or two berries would burst in my hand. I gobbled them up, licking the green seeds from my palm before grabbing another handful as their licorice-like flavor lingered between the roof of my mouth and my tongue.

The berry plants also grew in a sunny, sheltered corner next to the concrete front steps of the farmhouse, where they had re-seeded themselves from the raised flower beds above. The precise path which bore the berry seeds from the garden across the driveway into the flower beds was a secret shared between the garden and the prairie wind that visited it. Ants lived beside the front steps too, small granular hills marking the doorways into their underground subway system. Grandma and I were watching them. "Here,"

she said, "put this beside them, see what happens." Swept into her hand from our lunch sandwiches at the table inside, she dumped crumbs into my palm. I squatted, dropping the crumbs one at a time, aiming them into the path of an ant, as much as one can aim a crumb or anticipate an ant's path.

I observed an ant feeling out a crumb, then tried to see what little ant part it was using for dragging our lunch leftovers. I asked Grandma what kind of ants they were and she said we should look it up in the encyclopedia. When my legs got tired from squatting, I left the little village and went inside to the bookcase. I pulled out the "A" book, looked up "Ants." I carried the encyclopedia back outside and showed it to Grandma. That's when I learned they were pavement ants, that ants are very strong, that they have a queen.

The next time I came to visit, Grandma had a plastic ant farm waiting for me, two plexiglass sheets in a frame. We packed the space between the sheets with dirt, then collected ants and flicked them down onto it, snapping the lid closed. Sometimes I remained indoors, watching the ants shift one granule of dirt at a time in the see-through tunnels of this second subway. Other times I went outside to examine the ant village from above, poking stalks of dried grass into their holes to see what they would do.

When we had to leave the farm to return to Alberta, I cried and cried. I wanted Grandma, my cousins, my aunty who was only a few years older than me. I wanted to walk forever in the woods outside, seeking the carved initials of my aunts and uncles, a scavenger hunt marked by ridges of scarred wood. I loved exploring the collapsed granaries, lifting warped boards to look for mice nests, climbing over my aunty's childhood forts, buried in long green grasses. I wanted to keep searching for mica in the hidden valley, to slice between its soft rock layers with a sharp knife. I would miss eating the berries, I would miss the ants. I could hardly bear it.

When I was an adult, my husband and I lived in Saskatchewan for a while, a precious, golden, glowing return. By then the farm had been bought by one of my uncles, and Grandma had moved into a house in town. We didn't have a garden plot at our house, but she did: the empty lot on her property had grown veggies ten years ago. She and I decided it was time for it to produce again.

We discussed what we should plant where and whether we should run rows vertically or horizontally. We both knew she was the one with the real experience, and anything I offered was naïve, but she entertained my opinions anyway.

We planted peas from the Old Country. I didn't really know what the Old Country was—maybe Russia?—and questioned silently how the seeds could possibly still be pure when they'd had decades to cross breed. Still, I rolled a hard, wrinkled pea between my thumb and finger and imagined my great-great-grandmother doing the same.

Beans, onions, beets, lettuce, mixed flowers, and, of course, potatoes kept company with the peas. Each time I returned to the garden, Grandma had incorporated something new. She planted more flowers in all the available spaces, tucked between the rows. She hilled the potatoes, formed troughs for the beans, and staked the peas. When I watered, she taught me to point the hose a couple of inches away from the base of the plants so as not to shock them with the cold.

Seven months pregnant with my first child, I sat in the coarse grass at the edge of the garden with my knees up, reaching far over my belly to pull weeds. "You shouldn't be working so hard, little lady," Grandma said. I snorted. She had raised thirteen children of her own and six orphaned nephews while living on a farm with an alcoholic husband. As I pulled more weeds, she sat on the deck in her long pants and wide-brimmed gardening hat, facing me. She told me about the happenings in the lives of other people in town, as if I knew exactly who she was talking about. As she spoke, she paused once in a while to name the bird flutter-hopping from the top of one pine tree to the next, or the bird swooping overhead for insects.

I was stretching beyond my weed-pulling reach when I noticed the berry plant, then another and another. This patch of garden that hadn't been a garden for ten years—had, in fact, been grown-in lawn until seven weeks ago—was sprouting the berry plants from my youth.

When I finished weeding, I brewed us each a cup of tea and then sat beside Grandma on the deck. I pointed out the spider picking its way along the tightrope between two geraniums in her planter, and we kept track of its progress while we sipped and watched the sky paint itself into darkness.

As the berries ripened—the berries we mistakenly called blackberries as kids—I picked handful after handful, some for me, some for the pail. I told Grandma I was going to make jam although I never had before. Later, cooling jar in hand, I centered a label over it, then stuck it on, smoothing the edges. It was time to write the jam flavor and the date on the label, to be inducted into the club to which generations of women in my family belonged. That labeling was the official stamp of a jam maker—except I didn't know what kind of jam I'd actually made. I looked it up on the Internet, scrutinizing pictures and descriptions of berries, until I knew I was right: I picked the more magical name of the two the berries went by, the more appropriate name. I went back to my blank label: *Wonderberry Jam*, I wrote.

We planted the garden together the next season, and in the season after that, it was my son Grandma taught where to point the hose when watering. That fall, she sat on a chair in the garden beside tall wildflowers and held my swaddled infant daughter on her lap as I harvested the rest of the potatoes. My son was on his knees plucking seeds from a sunflower head, making hills of them on the soil at her feet. He paused now and then to grab a fistful of berries from the pail beside him, dirt sticking to the juice on his hands.

Two years later, having moved to the city but back in Saskatchewan to deal with the house we still owned there, Mom asked me to take some flowers to the hospital for Grandma, for her birthday. Grandma had been in and out of hospital for most of that time, and now the doctor thought her cancer might be back.

I peered around the corner as I stepped into Grandma's room, hesitant in case she already had a visitor. She was alone, lying on her side toward the door, blankets tugged up to her jawline. She wasn't wearing her glasses, might have been dozing. I walked over to the bed.

"Hello," she said, opening her eyes.

"Hi Gram, I brought you some flowers for your birthday, from all of us." I held them low in front of her bed, in her line of sight. "Can you see them?" I asked.

"Yes," she said smiling, but I wasn't sure.

I had visited her in hospital often before, had driven her there sometimes myself. She always wanted to make sure she had goodies to offer the people who came to visit her and would send me to the gift shop to buy butterscotch candies or a package of cookies. No matter whether she felt tired or ill, she never sent anyone away so she could rest. "Here, here, have a little goodie, help yourself," she would say from her hospital bed. "I'm in the best place, getting the best care," she would add, as a lab tech came in for another vial of blood or the nurse interrupted to adjust her IV. "That's fine, you don't have to leave," she'd tell family if they made moves to let her sleep. "So tell me about your day."

I set the bouquet I'd brought on the ledge under the window. There were two other vases of flowers there and a half dozen cards. I didn't see any candies or cookies.

I squatted in front of her beside the hospital bed and rested my hand on her shoulder. "There were some cedar waxwings outside our kitchen window this morning," I said, "eating the frozen crabapples on the tree." I could picture them, sleek gray and sharp black, fluttering wings knocking pomegranate fruits onto our driveway below, rolling piles to be swept away later. "I lifted Georgia up onto the counter so she could watch them. It was pretty neat." I bit the inside of my lower lip, hard. "I learned that from you, you know, appreciating nature."

She smiled, her eyes closed.

"Okay Gram, Happy Birthday, love you."

"Love you more," she said, eyes still closed.

I kissed her cheek, gave her shoulder a little rub through the blanket. I stood up and looked at her, drank her in. Left.

The hospital door delivered me into an alternate reality. My family was waiting in the idling van across the parking lot. I hefted my feelings into the passenger seat while the kids watched their DVD in the back and my husband texted on his work phone.

I turned toward the passenger side window, my back a shield, my mind still in the room with Grandma. "Are you okay?" my husband asked as we drove out of the parking lot. I couldn't quite answer him.

I looked out my window at the Saskatchewan prairie rolling by. It was the landscape of my childhood memories, cradling my first loves. The purple crocuses and squatting prickly pear cactuses, the white-tipped foxes and poking antler sheds. The berries, the ants. Grandma. A few of my tears overflowed their eyelid cups, my reflection superimposed over the fields as we passed by.

Four weeks later, I was in the van again when Mom phoned. Grandma was dying—they were calling the family in. Mom would be leaving within the half hour to make the three-hour drive. We hung up and I knew that with two children to organize and my husband at work in the city, I wouldn't make it on time. I sat in the parking lot. I thought about waxwings.

Now the wonderberry plants have grown, although late because I sowed the seeds too deeply at first: it was only after they should have sprouted and hadn't that I'd remembered Grandma saying not to push the seeds too deep, they need light, so I'd scratched the seeds closer to the surface. I see green berries hanging in clusters, but I don't think they will ripen in time. I check them every day anyway, searching for purple. Then two hard frosts curl the plant leaves over the berry bunches and I sigh at having to wait another whole year before I can taste the Saskatchewan prairie again. The frosts are followed by two weeks of hot summer weather and I feel robbed because my plants would still be full and vibrant if not for those two misplaced nights of silvery cold.

One day, the kids and I arrive home after I've picked them up at school. Moving into the backyard from the driveway, I see shiny red strawberries in our berry patch and call the kids over. They squat to pick them and I trail my fingers through the curled wonderberry leaves. Then I see purple. I am lucky to grab a few berries as the kids, alerted by my exclamation, swarm me and scoop the rest, licking the seeds from their palms. ❧

Dusk

The rabbits here are swimmers,
spend their days submerged
in shallows, just ears and eyes
breaking the surface of brackish
water ringed in sea grass.

Sometimes, when I'm wading,
fur brushes my legs, a soft echo
of my mother's white muff,
lined in white velvet, a richness
I stole from her closet and kept.

Come evening, you can measure
distance in the round, quiet bodies
of marsh rabbits come ashore:
brown nails studding the green grass,
fastening the ground to the road,

the road itself anchored to August
by a moon doused in brine.

Reservoir

We lower our July bodies
to immersion, some shell
of the world breaking open
to water, quiet in a way oceans
cannot grasp, a saltless float.

Here legs cut through to colder
layers, hair dreads, pale skin
flashes the language of must,
will, and watch me. Everything,
finally, a test of buoyancy.

MILLING

We tumbled our apples—stems, seeds, and all—into the wooden hopper, where they were snagged by iron tines, pulled through blades, and shredded into a sweet pomace. Collected overripe (best for cider), the apples matured now into their cider lives. We milled them and recalled our first encounters with nature, memories sweetly emboldened by youth and muddled in newness.

Ice Walking | FICTION

WHEN I FELL THROUGH THE ICE, time stopped. The light faded, water enveloped my body like a slick, sharp cocoon, and that fleeting mystery of life, always one step ahead as we barrel headlong into the future, was suddenly thrust in front of me. So *that's* what it's all about. Sinking into the blackness, wishing I had learned to swim so many years ago, wishing my mother had taught me, I saw the meaning of my life. The signature, the crux, the pinnacle on which everything now balanced. Swimming. You should have known all there was to know about *swimming.*

I was afraid of the lake in my younger days, fearful that an ancient, slimy hand would latch onto my ankle as soon as I wandered out over my waist, pulling me down into some rocky crypt. The terror seemed so real and, though that razor-sharp certainty dulled as I grew older, I never stepped back into the water. Swimming was a choice, after all. All the other kids were swimming, but they *wanted* to. My feet were better suited to dry land, even though our family had access to an acre of marvelous lakefront property in Bridgeton, Maine. A great deal of my life was centered around the family camp, around the lake, but simply gazing over the water was enough for me. I remember practicing my breathing skills in case I was ever thrust into that murky keep. One time I held my breath for two and a half minutes. My father told me that was on par with the Navy SEALs. I only shrugged.

As winter descended again and again, and I grew from a child to an adolescent, I began to wonder what stepping onto the ice might feel like. My father would set ice-fishing traps from time to time, but he knew better than to invite me. He'd glance at me uneasily while loading cans of beer into his pack basket, watching as I stared out over the lake, bright white in the low winter sun. He was probably close to asking me along on a few occasions, the words sitting on his dry tongue. And then the cabin door would slam, and I would watch as he rode off on his snowmobile over the ice, the wild cold inviting him in.

Maybe I'd try a step, I thought one Saturday in mid-January, another average weekend at the cabin come round again. Just a step, and then I'd go back. The night before, Dad had told me twelve inches of ice could hold a Mack Truck, and twenty could hold a castle. "Stronger than granite, boy," he said as my mother looked on, emotionless. The three of us sat around the dinner table, picking at our food as the blackness of space pressed in through the cabin windows. My father was eating in great gulps, chewing little, taking a slug of beer every so often. I wanted to ask him how he knew those things about ice. Years later, I realized that wasn't the point. When Dad said something, it became fact. It was right, no matter what. Mom and I both knew it wasn't worth questioning him. Anger was his strong suit, right up until the day he died.

So I put on my snowsuit that day, a boy of fourteen just beginning to discover the great irony of life, that our parents know as little as us, and trudged out into the frozen wasteland. My mother was still sleeping, as was her custom. The first few breaths of cold air caught in my throat and I coughed the air back out. I zipped the suit up all the way so only the tip of my nose was exposed. My eyes were watering by the time I reached the lake. I scanned the horizon in search of my father, who'd departed at daybreak to "find the honey hole." The far side of the lake was three miles across, hazy in the white sunlight that danced above the ice. To my left and right, the lake seemed to stretch on forever. Its size had always frightened me, but today I just thought of Dad, out there in the middle somewhere, his snowmobile purring, that mystical frozen fortress in his sights.

I peered back up at the cabin, part of me hoping that Mom had awoken, that she would call out and stop me, bid me back into the warmth. The sun's reflection was harsh and I couldn't see through the windows.

I turned around and took my first step onto the ice. Snow was packed tightly on the surface, and my foot sunk in only slightly. I brought my other foot down, and there I was, standing on the lake, defying all the laws of my life. I stood there for a few moments, the cold air pushing in on me from all sides. What happens next? I wondered. Is there anything else, anything else in the world now? Facing great fears always conjures up more questions: what's left to dread? The water was still under the ice, of course, just like the stars are still in the sky on a bluebird day. But the stars can't get us, I thought, and neither can the water. So I started forward.

I knew my father liked to fish the far side of the lake, and now that I was out here I felt the entire sheet of ice was mine to dictate. I would traverse the lake, my snow gear thick and warm, fully intact. Plenty of sun left in the January sky. I would surprise that bastard for the first time in my life.

I passed by a small island where ice flows had buckled up to the tree line. A red squirrel chattered out at me from one of the low spruce branches. I chattered back, the sound startling and grand. I was an explorer now, finally, at fourteen. I watched two chickadees dart through the underbrush, a game of chase most birds were now playing in the South.

Running now, skipping, I put the island behind me. My boots crunched into the snowpack, the frozen lake a vast field wrapping in all directions. I was growing accustomed to the cold air, cherishing its freshness. My nose was red and sore, but what did that matter? I was an *explorer*.

At some point, I had to look back to get my bearings. The other side of the lake wasn't getting any bigger on the horizon, and for an instant I wondered if I was the recipient of some cruel prank. As I turned my head to look—the cabin was at least a mile to my rear, now the size of a pea—my boots tangled together and I went sprawling forward. I hit the snowpack with a thud, and my chin bounced off the cold, jagged surface. I took off my gloves and brought a hand to my face. Surely there would be blood. But the skin was numb, my hand dry. I started rubbing my chin to try

and get the feeling back. That was when I heard the snowmobile.

Dad! I thought instantly. But when I looked out across the lake, I saw four black dots instead of one. They were racing across the ice. The purring sound got louder, rising, rising, sawing now. I covered my ears as the snowmobiles approached, burying my head in my snowsuit, wishing they would leave me be. I only wanted to see my dad.

The sleds came to a halt a few feet in front of me, and one of the men hopped down from his seat and ambled over to me.

"You OK, son?" he asked, kneeling at my side. His Maine accent was deep and impressionable. He had a big frosty beard and dark matted hair. He wasn't wearing a helmet. "What're you doin' way out here?"

"Walking," I said. "Exploring. Looking for my dad."

The man stared at me for a moment, and he lost something in his eyes.

"You Tim Donaldson's boy?"

I didn't have to say anything. He knew the answer.

"Let's get you back to your mother. There's been an accident."

Everything in my life happened quickly after that moment. Scenes went by in flashes: the man lifting me onto his snowmobile with iron hands, the sharp gnaw of the wind on my face as we raced back to the cabin, my mother's tired eyes, the lack of tears there, my aunt and uncle arriving, the strange men in suits, the talk of open water out on the lake somewhere, that this winter—despite the current cold spell—was going to be the warmest on record, scuba divers, one hundred feet of water, nothing turning up, one of my friends calling to say he was sorry, the rigid funeral service. And then, emptiness.

That March, my mother and I went back to the cabin. I don't know why she wanted to take me, or why she wanted to subject herself to my father's specter. She mumbled a few things about packing up, making the proper arrangements, enjoying the place while we had it. She was putting the cabin and the acre of land

Spawning III (Upstream) | JENNY KENDLER
Graphite, iridescent ink, and colored pencil on paper, 11 1/4 x 15 in., 2009

on the market come spring. "And whatever you do, stay off the ice. You can't even swim. Will you ever learn, Nick?"

"It's too late to learn," I replied, closing up a box of winter clothes as the sun set over the lake. I stared for a moment but the glare was too bright, so I looked away.

My mother, so mousy and lifeless, smiled sadly. "Your father knew how."

"I'm going to bed."

As I lay awake that night, the quiet of the wild deafening, I decided it was time. I went to the closet and bundled myself in the proper attire, and then crept out into the hallway. If I woke my mother, the game was over. No shifting floorboards.

After sliding on my boots, I stepped out into the deep, black night. The air was warm for March, unseasonable, but the cold clutches of winter still stuck around in the shadows. In the ice. The shal-

low parts of the lake were covered in a twelve-inch layer—*that Mack Truck's comin', boy*—though the consistency was soft. I could have drilled down and seen the purplish hue, the layers of bubbles frozen for a winter's time. But I had no drill, and I had no care anymore. All I wanted was to finish the walk I'd started back in January.

So onto the ice I went, nothing but darkness ahead. My boots sank into the ankle-deep slush, every step a chore. The air hinted at an early spring, all of those smells of the wild awakening: lavender, the moisture of shadows, green moss, time. I slogged through the soup, that wet crunch the only sound in the world.

Eventually I stopped and listened, wiping the sweat off my brow. Overdressed, I thought. Somewhere across the lake—or maybe it was everywhere—the ice buckled. The sound was like a great cable snapping in two, echoing as the pieces recoiled, on and on, reverberating like feedback. My footing shifted. Fear spread to my heart, that great truth of the water's mastery over us. But I didn't fall through.

I continued headlong into the night. After a time, I stopped again. In the distance I heard sloshing. My mother? Surely not. She would be screaming my name, her footfalls messy and uneven. This sound was more of a scraping, a clockwork shifting. Soon two yellow eyes appeared in the gloom some twenty yards to my right. The shifting stopped, and the eyes waited. Nature was patient.

I started up again, glancing to my right ever so often. Whatever creature the eyes belonged to was following me. A companion. I smiled; I had never had a companion before.

I don't know how long I walked. Time seemed to shrink away, and I was left with something else. A voice inside my head told me to go back. Somewhere up inside.

Run, said the voice. *Run home, Nicholas.*

So I ran. But not home. I'd come out here for a reason, after all. Slush and standing water splashed about, soaking my snow pants and spraying my hands, my face. My companion followed. I thought I heard a yipping noise. My heart beat out of my chest as I ran, ran, ran.

Then, in an instant, the sky brightened. The expanse was now cobalt gray, the color of housefire smoke, and the outlines of trees on the horizon merged into grainy focus. The trees went on and on, north and south.

Another *yip-yip!* My companion's bright eyes faded some, and what appeared now was a scrawny coyote. His fur was thin and soiled. He'd fallen fifty yards behind and was standing in a pool of water. He yipped again, and I stopped running.

"Come on, boy!" I called back to him. "It's OK! Just a little slush. I'm headed to find the ice fortress!"

He yipped again, a high-pitched squeal that I'd only ever heard from afar, in the depths of night when all the coyotes hash over coyote affairs. He was looking past me, further on, so I turned to look, too. Had there been an ice fortress out there, once? Maybe. The coming dawn illuminated the open water just enough to differentiate it from the slush-topped ice. If the sun had been out, the water would have shimmered. But today the surface held a non-quality.

Dad's world, I thought, and I was going to turn back, but before I could pivot my feet, the ice gave way.

A sickening grinding noise crackled through the still morning air, and for a moment I was weightless. Then I was in the water. The cold wasn't what got to me. Or the thought of my father calling out as he went under. What got me was the *swimming*. This was why I had avoided the lake. Everything was crystal clear now.

Oh, but my years of breathing exercises weren't for nothing. Before submerging, I took in a great breath and, as I sank, my panic subsided. My body became numb, but my eyes worked just fine. And though there was nothing to see, the darkness was beautiful. The air sat comfortably in my lungs, and I decided I would go for a new record. Three minutes.

At some point during the sinking—one hundred feet down or so—a fish swam by. I don't know what kind. A bass, maybe. He circled me twice, and then nibbled at my right hand. *Ouch*, I thought. *Stop that. Wait till I'm dead.*

Bugger that, the fish said. *No dying for you. Too much to do, yet. Besides, this is my lake.*

I'll never make it back up.

The fish laughed.

I don't know how to swim.

Everyone knows how, boy.

I thought for a moment. I was still sinking, and the fish was following me lazily downward.

Are you my father? I couldn't speak. My precious air would vanish, and that would be that.

The fish grinned, and I thought I could see my father's grin inside him.

The fish circled me and nibbled my finger again.

I pulled my hand inward, and as I did so, I felt a little lighter. I moved my hand again, and then both hands.

That's the ticket! the fish hollered, circling excitedly.

The trouble was, I was beginning to like it down here. Even if I do get back to the top, I'll have hypothermia. No one's coming. I'll freeze.

You'll be up there, though. Isn't that enough?

I don't know, I thought. But the fish was gone.

I started kicking and, exhaling, I followed the air bubbles upward. ❧

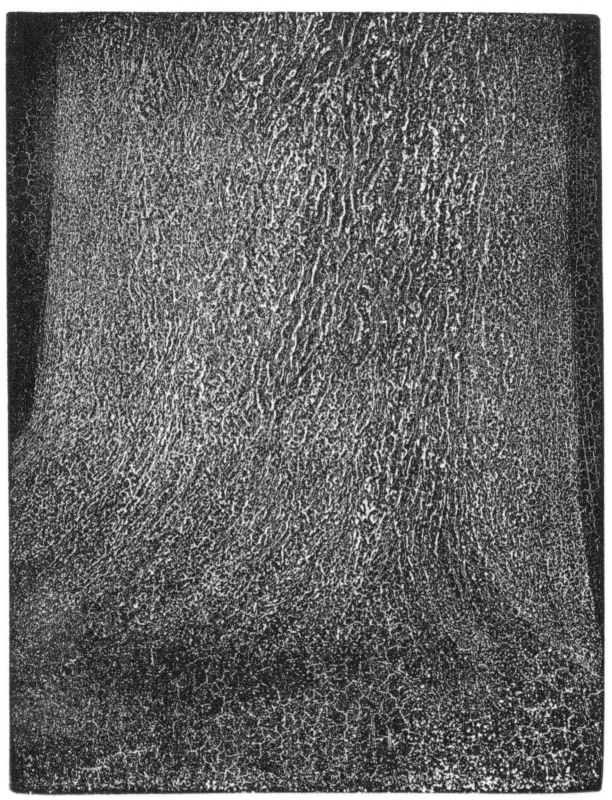

Tree Trunk I (Willow) | BRIAN D. COHEN
Relief etching with watercolor, 8 x 6 in., 2011

Tree Trunk II (Douglas Fir) | BRIAN D. COHEN
Etching, 9 x 6 in., 2015

Tree Trunk III (Douglas Fir) | BRIAN D. COHEN
Etching, 9 x 6 in., 2015

Tree Trunk IV (Maple) | BRIAN D. COHEN
Etching, 9 x 6 in., 2015

Pastoral

The ewes begin to stir when they see us coming.
From far across the field, leopard-spotted
with half-melted snow and tufts
of tea-colored grass, we hear the slush
of a dying winter

and as we near, we see the pink
spring petals of their tongues.
I've brought my young friend,
three years old, to see the lambs.

No, she doesn't want to touch.
She must feel, I think, the meanness
of straw beds beaded with feces,
of bottles of soy milk that give stones,

though she hasn't asked about the dead one.
And perhaps she feels, as she stands there,
her face to the blustery air, what
the long cold nights must be like—

the unfathomable snowfall as
lambs burrow into the fat sour warmth
of a mother ewe, who keeps one eye open,
waiting for the dawn.

KRISTEN M. PLOETZ

When the Cardinal Takes Flight | NONFICTION

"**D**O YOU HEAR THAT?**"** I whisper to my daughter. "It's the cardinal," I tell her.

She stops what she's doing to find him, her lake-blue eyes scanning the lone, leafed-out maple in our yard. His red cloak is camouflaged among the fluttering leaves. Elusive and coy, he wishes to remain unseen.

"I can't see him. Where, Mama?" She turns back to face me.

I take in the full picture of my soon-to-be eight-year-old: gap-toothed innocence, boundless curiosity, and a haircut that's long overdue. Before I can answer, the cardinal emerges in full and furious flight toward the neighbor's oak tree. A new vantage point was needed. He finds a suitable perch and picks up where he left off.

Birdie birdie birdie birdie.

Not much in life is certain. Yet I trust my daughter will learn there is beauty in this world, even if it is not always visible. The cardinal's call will remain steady, just as it has for the time I've come to know it. Like I have, she will come to understand some things are reliable. I am consoled by the knowledge that she can always listen for his lyrics, even when I am long gone. I don't have religion or recipes to pass down. There are no guarantees I can give about paths to future passions or professions. Only this.

Later, my mind wanders while I dig a shallow trench for the Purple Majesty and Yukon Gold. Thoughts of potato salad at late summer cookouts keep me motivated in the scorching midday sun. Then I spot it: a slender ribbon of soil undulates just out of reach.

"Oh, look how big this one is!" My excited utterance is involuntary, as it often is. I am usually by myself, but on this day she plays nearby.

She extends her open palm toward me. "Can I hold it?"

I pluck the earthworm from the freshly dug hole and drop it in her hand. Wriggling induces giggling.

Most of her friends would not willingly partake in such delight. I am uncertain whether I should be proud of her or saddened by this modern state of affairs. Perhaps both.

We talk about compost and castings and their particular arc along the circle of life. She already knows death and decay are a necessary corollary to life, but also that it possesses its own kind of heartbreaking beauty. Still, there's no reason to have it unduly hastened. She sidles down the garden path a few steps, and after a brief (arguably one-sided) conversation held at eye level, finds him another spot in the soil, safe from my garden spade.

It's the end of another day, one filled with errands and homework. If I'm not careful, Banality too frequently takes over as puppet master of our days. Oh, how she deftly works the marionette strings with an unappreciative ennui, relying on predictable endings that offer solace and satisfaction, but little surprise. Her calloused hands don't always feel the tension in the strings as she tries to maneuver through yet another day like this. It remains unclear whether this makes her my muse or simply a menace. But tonight I quickstep backstage. I demand that Banality hand over the controls. She relents with a smile.

"Quick! Come on the deck. Look at this sunset. There is so much pink tonight!" I shout through an open window.

Bare feet scamper across the wood floors. The kitchen door slams behind her.

"Wow," she sighs.

For a moment, we are connected through our shared awe of this scattering light in the twilight sky. We watch the colors shift. Cotton-candy pink yields to deep, dusky violet. Gunmetal creeps on hindquarters toward black, and the margin of evening has kissed the day good night. The kaleidoscope of colors may vary in opulence, or even become occluded or obscured from time to time, but the close of every single day offers its own kind of consolation and countenance. She already expects the pattern to repeat again tomorrow and the day after that.

A hush falls over the theater. We are still in the backyard, and the final act—bedtime—is upon us. Pajamas, books, and good night are no longer met with resistance. It is now, of course, an easy path. After this many years, its terrain is expected and familiar, for both of us. I indulge in this reprieve once it arrives each evening—I know what waits for me on the other side of her slumber. Yet, as the darkness slowly fills in around us like a stub of charcoal quietly rubbing against cotton bond, filling in from edge to center, we linger.

"Remember, look for the three stars in a row. That's his belt," I whisper into my daughter's ear, her flaxen hair brushing against my lips.

"I see him. Right there, Mama!" she exclaims with delight, pulling me tighter to her lanky body.

As we tilt our heads toward wonder and away from the clock, I remember one of her first words: moon. "Moo-in," she'd say, pointing her pudgy finger at the gray-white slice of rock smiling back at us. A few years later she learned the difference between the Corn Moon and the Worm Moon. Last summer she danced on a sandy Rhode Island shore, frolicking before the swollen pink supermoon that hovered low above the Atlantic, perfectly punctuating the end of our vacation. These are the primitive teaching tools offered by the night sky. I use them to show her there is beauty beyond us, and we can count on it to offer context for the deepest of loves. Complexities found in well-charted constellations and the orbits of comets will soon follow. There's a new cadence to her understanding. I sense that it is quickening, deepening.

When life is lived the right way, with vulnerability and bravery tendered at the fore, subtle shifts and seismic faults will inevitably etch new contours in the heart's landscape. The bedrocks of love and friendship might be discovered as unreliable, unrequited,

or else in finite supply. People will move on without you—whether it is deliberate or demanded by death, in both cases it hurts. New maps are always needed to navigate the shifting terrain, to recalibrate the cardinal directions.

My daughter has not yet lived long enough to learn these truths. I want her to be prepared. I want her to become familiar with what she can trust. When so much will be beyond her control, I want her to find solace in the wondrous, beautiful things that remain constant.

My outdoor classroom is small for these lessons on the fly. We do not have mountain vistas. There is no roar of the ocean at our back door. Homes in this dense neighborhood are the same age as her grandparents, sitting mere feet from the property line. Yards are measured in square feet rather than acres. It can feel crowded, if you let it. I must compete with the din of the highway traffic and overhead jets painting the white noise around our days. The trees are old and sparse, surviving each season of Nor'easters with another hard-won ring in the sapwood, but fewer branches. Skittish and scrappy, the resident wildlife offers few opportunities for sustained awe.

Still, I find ways to make our world feel vast and steady, a place of comfort and quiet wonder. I help her knit a chain mail of resiliency by showing her how to pay attention to the world around us. The familiar flicker of a blue-feathered wing. Thousands of helicopters that drop each year from our maple tree, decorating the yard like confetti after a raucous party. A roly poly bug that hugs itself into a tiny armored sphere when she tries to coax it into her hand. I want these constants of our world to run along the bias, bending with her when she tumbles through the heartaches of life. Sunsets and stars, seasons and stones, songs and stamens—these will remain steadfast when all else falters.

And so we dig. We listen. We look up. We outstretch our hands. We open our hearts. ❧

The Starfish

As long as they both lived,
my parents told of the starfish
they beachcombed in California
and triple-bagged
for the two-day train ride home.

Laid out in the Illinois backyard
to air off before cleaning
and a lacquered afterlife,
the carcass basked
until a possum seized
one spiny leg, like any other scrap,
and, climbing up and over
a redwood fence, took all five
from future mantelpiece to legend.

Questions on Toads

Where do they begin
and the mud around them end?

What wand was waved
to animate the dust just
long enough to sculpt it,
leaving lumps and bulges to jump
and—when they must—couple,
hunt by luck and tongue
but mostly sit,
little more than mineral
under twigs and duff,
among protruding roots and rocks?

If clusters of protein and moisture
are what these warted
forms amount to,
why is coming upon
a plane of skin and guts
crushed under a truck tire
(or, if young, underfoot)
attended by, however small, regret?

PRESSING

The pomace was ripe and tangy and rested in the barrel like applesauce, chunks of fruit buried in globs of white pulp. We set the press plate on the mash and began to crank, wringing juice from the fruit like water from a towel. Amber liquid poured from the spout. We thought about the earth's gifts, about offerings as small and sweet as scarlet apples, pushing, pressing to know our place in this world of wasp and apple and carbon climate.

Food for Bears | NONFICTION

AFTER THE AUTUMNAL EQUINOX, the days grow quickly cooler. We go outside one night to view the eclipse of the perigree moon with its strange, orange light. I don't go outside at night as incautiously as I used to do. We're accustomed to owls, bats, raccoons, and coyotes accompanying our nighttime forays on the farm. After this season's encounters with bears on our land, I'm more aware than ever that I'm not always alone in the dark.

I spend the night after the equinox at my daughter's house, waking just before light to wait for my three-year-old grandson, who wakes up early himself. In the hallway nightlight I see his silhouette appear round the corner of my doorway as he pauses to see if I'm up. "Good morning," we say at the same time and laugh. As he pulls himself onto the bed, I ask, "Did you have any dreams?" I'm not sure he knows about dreaming yet, but he nods.

"I dreamed about Charlie," he tells me with a serious look in his eyes.

I haven't heard about Charlie before. I wonder if that's the name of a kid from school. "Who's Charlie?"

"He's a friendly bear, but he has a loud ROAR!" (The last word is delivered just as you'd expect from a friendly-but-loud bear named Charlie.)

"I'm glad you have a bear friend named Charlie," I tell my grandson with a hug. At three, his world is filled with animals from picture books, the popular educational video series *Wild Kratts*, and frequent visits to the zoo. I'm especially glad that my worries about this summer's bears haven't deterred him from imagining one as a friend. If I didn't know it before, I'm certain now that bears walk alongside humans, an arrangement that benefits us both. Even though they're wild creatures, bears can be our companions on this earth. I hope my grandson will always feel friendship for bears.

"Weather whiplash" is the term I've heard lately for the increased unpredictability and sudden, ping-pong changes in weather patterns these days. Weather has always been unpredictable for farmers, but the changing climate makes it even more volatile as weather patterns now veer toward extremes. Having a lifetime of familiarity with the weather in this region, I know we've entered an era of uncharted climate conversion, but to what we're converting isn't clear. A three-hour hailstorm—an unusual event on our farm—destroyed the entire crop of tomatoes we'd planted just that day. A long, warm autumn the year before seemed like a good thing when we were picking tomatoes at the end of October, but was devastating when the first freeze, a week into November, brought a sudden, nearly eighty-degree drop in temperatures, shocking trees that hadn't yet begun to go dormant and killing 30 percent of our fruit trees, some of which had been standing for more than fifty years. Because farming operates in seasonal cycles, that abrupt freeze meant the loss of next summer's fruit, with no apples for gallons and gallons of cider like we'd just pressed a couple months before.

Halfway through June, I am eating lunch with my partner, John, on the screened porch of our farm's community room when we sense something moving nearby. We are not at all prepared to see a magnificent black bear with a tan face amble around the corner of the ditch bank and onto the wooden bridge fifty feet from where we sit. We watch in stunned silence as the bear sits down, Buddha-like, on the planks near the end of the bridge, licks its paw, and looks around.

"What shall we do?" I whisper without turning my head.

"Nothing," John answers. We stay as still as possible, intent on the bear's every move as it sits serenely on our wooden bridge, watching the water flow. It doesn't seem to see us through the screen, but it may smell us or something of the human about our place. Soon it puts its front paws down, turns around, and wanders off the way it came, stopping to tip the nearby bench with its paw first in case it finds food underneath.

Did we really see a bear? A *real* bear? It came and went so quickly, it seemed more a spirit than a wild animal. Still, I wait a few minutes before tiptoeing out across the bridge to check where the bear has gone. Bears can move very quickly; by the time I cross the bridge, it has disappeared into the trees along the ditch. It occurs to me I've been enchanted by the bear's wild beauty, lustrous fur, and wise face. This desire to follow must be what "animal magnetism" means. Even though it wasn't doing any harm on the bridge and didn't threaten us in any way, I realize I'm not being prudent by stalking a bear. I want to see the bear again—from a distance—but I decide to give it a little more time to make its getaway, even if that means I will miss another encounter.

After fifteen minutes, John and I cross the bridge together and walk out to check on the bees in the east field. In hindsight, we might have taken the truck for protection, but we were still under the spell of the encounter. Holding hands and looking into the ditch for anywhere a bear might be hiding, we find only a trampled spot along the bank where the bear stopped to check for bugs at the base of some trees. We don't run into the bear again and the bees are fine, but the next day we install an electric fence around the hives, just in case.

I've had near-bear experiences before, but none that prepared me for so large an animal at so close a range. From piles of scat and paw prints to slashed melons and mangled beehives, we've found evidence of bears on our land and John has seen one here from further away. Still, we were surprised to see a bear in the middle of a June day; they usually come down in the fall before hibernation.

What we hadn't taken into account was last November's freeze coupled with a late spring frost. The same seventy-some-degree drop in one-day temperature that destroyed our fruit harvest also decimated the food supply in the mountains that bears and other animals depend on, especially as they prepare for their long winter hibernation. Biologists call this phenomenon "food collapse," as the food chain loses one of its links. Weather whiplash strikes again.

Later that night, we hear a noise outside like a door slamming. Ten minutes after that, our neighbor calls to say the bear was in her yard and is now heading for the highway. I run down our driveway with my camera in the hope of getting a picture from a safe distance, but when I notice that the bear has knocked over our trash can, I think better of being outside with an animal that large roaming around.

I call my daughter about the next day's visit to the farm with my grandson. We decide we can keep an eye open for another encounter while still having fun outside. We're on the lookout, especially when we notice a car slow down on the highway as if an animal might be crossing, but we don't see a bear that day. I am both relieved and disappointed not to share the beauty of such an animal with my grandson.

Like humans, bears are opportunistic omnivores. They'll eat what they find. When encountering a bear, it's good to remember that, with the exception of grizzlies, humans aren't food for bears; we are not their prey. They'll only attack if they feel threatened. Keeping a safe distance from all bears is the best plan, even if we're drawn to their charisma. In my enchantment with bears I had to remember that their claws—not to mention the muscle and bone behind them—can easily maim and kill.

As the summer days pass, a certain kind of dark shadow in the trees makes me pause. If one bear has come down from the foothills, what's to stop another from following? On the way out to the garden, we find a small paw print in the mud, probably a young bear checking on the bees, which we're relieved to see are still safe. Next we find a large pile of oily bear scat near our greenhouse, a little close for comfort. And then we find the back barn door pulled out from the bottom as if a bear has tried to crawl underneath.

Worried about our members who come to the farm for their share of vegetables each week, we notify them that bears are in the area and put strong latches on the barn and chicken coop. We caution everyone to close the doors behind them when they're in the barn and when they leave. No one complains; our members are sympathetic to the bears' plight. "Poor bears," they say in the barn. "They're just hungry." Despite this empathy, we hope that none of our members will run into a bear, just as we hope that a bear won't run into one of our members.

Overset | CAROLINE MILLER
Digital photography, 2015

The bears aren't just at our farm. They're spotted in town, crossing the river and wandering through people's backyards. With so many bears around, I start to anticipate another bear encounter. A friend who lived in Alaska teaches me what to do if I confront a bear: clap my hands and yell as loudly as possible, "Go away, bear! Go away!" Another friend suggests banging pie plates together when we're outside at night. I don't go quite that far in precaution but I do practice clapping and yelling, just in case.

And so the season goes. We make the summer's first pesto, cover our crops with net to deter deer, and hope the second round of tomatoes has time to ripen before the first fall frost. We pick rhubarb in July, something we've never done before, since it's usually an early season crop—and that's weeks after the hailstorm ripped the plants to shreds. Rhubarb revival, I call it. We are happy to have more rhubarb, but it's unsettling to realize our climate has changed enough to alter the growth pattern of a perennial plant. Perhaps the hail stimulated the plants into going to seed again as a survival mechanism. Is rhubarb sending us a lesson about adaptation we ought to heed?

I read a report about governors in states with large rural populations meeting to discuss the impact of climate change. People in rural areas, they realize, will be more heavily impacted than people in cities, at least initially, since we depend on weather for our livelihoods, live closer to the natural world, and have reduced access to emergency services. I never hear about the outcome of that meeting, but I am glad that officials are recognizing the difficulties farmers and others in non-urban communities are already facing.

Weather has always been the factor least under a farmer's control. Today, that incapacity is magnified by a political paralysis to stop the conditions creating even more instability in the climate upon which we depend. In the midst of all this uncertainty, one thing's for sure: it'll take more than banging a couple of pie plates together to face off what's coming.

At 7:30 on a Friday morning in August, our elderly neighbor calls for help. A bear has just attacked her goats, leaving one goat dying with a slash through its neck. The latch on her goat shed was broken and the goats had gotten out in the night, attracting a hungry bear looking for whatever food it could find.

She's scared the bear off for now, but needs help with the goat. As John heads over to see what he can do, I go outside to check the chickens. They don't seem disturbed, but as I'm standing at the coop with the hose in my hand, something tells me to look around. A very large black bear is approaching the chickens—and me—not more than twenty feet away.

Now is the moment for which I'd practiced all summer. I hang up the hose, clap my hands, and say in the loudest voice I can muster, "Get away, bear! Get away!" The bear pauses and hunches forward on its front paws, then quickly springs and wheels in a run toward the front of the house.

When a bear's muscles move, its fur glistens and ripples like velvet grass in a breeze. I pause a minute to be sure I've actually seen a bear and another moment to consider how it shook its big head as it crouched and turned like a kid with a dare: "Come and get me." Was this bear playing with me? Figuring I'm safer inside than out, I run toward the back of the house, realizing that if the bear circles around the bunkhouse, I may meet it on our back porch.

But I don't. I run into the house and down to the front windows to see if the bear comes by. No bear in sight. I call our neighbor to warn John that a bear is at the farm so he'll watch for it when he comes home. But when my neighbor answers, she says the bear must have turned again while I was getting to the house; they've just watched it run through their front yard and into the neighbor's on the other side, where it headed towards the foothills. For now, the bear has left the area.

During all this excitement, the veterinarian has arrived to euthanize the dying goat. John will come home for the tractor so he can transport the dead goat to our driveway, where animal cremation services can pick it up. I wait with our neighbor and the goat. I'm sad that the goat died but I'm still under the bear's spell—I can't feel as sad for the goat as I do for the bear, who had to leave its mountain home to find food down below. If it weren't for weather whiplash, that bear would never have killed that goat.

Our neighbor has already called the Office of Wildlife Management. After the officer talks with her, he comes over to ask if he can trap the bear on our property. We are really sorry to hear this; it's not a decision we want to make. We reluctantly agree because we think the bear will be back for other livestock—maybe even ours. With so many people coming and going here, we are nervous about anyone running into a potentially aggressive bear.

As I watch, the officer drives the trap to the back wooded spot between our two properties. He baits it with a maple doughnut, cantaloupe, canned cat food, and peach pie doused with anise oil, which apparently is attractive to bears. The trap is a six-by-four metal cage on a trailer. When a bear crawls in, its weight trips the heavy door shut. John and I check the trap before we go to bed that night but find it still empty.

My first thought on waking Saturday is the bear. We walk outside and can see it in the trap as we look through our binoculars from a safe distance. Even from there, we can tell it's a big bear from the way its silhouette fills the cage. We call the officer and arrange for him to come at 9:15, after our working members have gone out to the field. When they arrive, we gather at the barn to tell them about the bear, but no one goes back to the trees to see it.

I wait for the officer's arrival before I approach the trap. I feel sick to see the poor bear there with only enough room to turn around or lay on the floor, which it does with its sad eyes looking up at us as it cries. "Cry" is the right word; the sound is as heart-rending as the cry of our own species. "They make that sound when they know they're defeated," the officer tells me. Later he says, "This is the worst part of my job."

I watch as the officer hitches the trailer to the truck. While he is looking for tie-downs, I stand only two feet from the bear as it tries to bite the steel bar of the trap with its startlingly large teeth. "I'm sorry, bear," I whisper. For a crazy second, I imagine opening the door to let the bear escape. Then I realize that an angry, frustrated bear might attack the first thing it sees—me. I wonder what might have happened if we'd let the bear go. I also wonder whether we'd caught the right bear. While the size seems right, its muzzle seems darker than the one I'd seen the day before. The officer says it was most likely the same bear coming back for the other goats. He also says it was one of the biggest bears they've ever caught.

The bear keeps moaning from deep in its throat, a low, sad keening that I will never forget. With its huge arm, it hits the floor of the cage, startling the officer as he double-checks the lock on the door. "It's okay, buddy," he says as he puts a cover with air vents over the trap for the bear's trip down the road. The officer tells me he's angry that people don't take better precautions to keep the bears away from their homes, like locking up pets and livestock and keeping their trash in their garage. I know the saying "A fed bear is a dead bear," but I also know that in our changing climate, we can't always predict what a hungry bear will do.

I ask the officer what will happen next. They will take the bear to the wildlife office, give it something to knock it out, and then kill it. I don't think "kill it" were his words, but that's what he meant. "Can't it be relocated?" I ask. No, once a bear kills livestock—and they suspect this was the bear that had killed chickens in our area, too—it has to be put down: something to do with liability, the world having come to decisions made in this way. I ask when the bears will return to the mountains to hibernate. "Usually, when they're full," he says. "This year, we don't know."

I watch the trailer pulling away with its bear cargo and feel sick at heart. At least it was an older bear, the officer had said. It likely wouldn't survive the coming winter. I think that one less bear competing for food might be good, but I know that's wrong, a reflection of how out-of-balance our climate and its many ecosystems have become.

Now in the cool autumn, I scan the ground for prints or scat and check the compost pile every morning. After we saw the first bear, I stopped putting fruit scraps in the compost. Instead, I threw our rotten peaches and watermelon rinds in the irrigation ditch and watched them bob downstream. With no berries growing in the mountains, any whiff of fruit could attract a prowling bear. Since the officer took the bear away, watermelon rinds in the compost pile lay softly rotting.

I admit I miss finding evidence of the bear on the property; I miss the bear's presence. But I'd received our neighbor's early morning call. I'd scared the bear away as it approached our chickens. I'd given permission for the trap on our land. I'd stood with the caged bear as it struck the bars with its paw. I'd listened to it cry in admission of defeat. I'd watched the wildlife officer drive off with it in his trap. I'd learned a bear that killed livestock would be eliminated. I'd helped to kill that bear, and now it wouldn't be back.

If the latch hadn't been broken, if the goats hadn't been out, if the bear hadn't killed the goat. Going back further, if the November freeze hadn't killed the fruit, if the mountains hadn't suffered food collapse, if the climate weren't unstable, if we could stop ruining our environment. If, if, if. Sometimes I wake at night with my heart pounding, but not from fear of a bear. Instead, it is wonder I feel at having been in the presence of such an animal—and sadness to be a part of its demise.

On Christmas Eve at my parents' house near an open space across which wildlife still travels, we hear the yap of coyotes caroling. My sister carries my grandson outside to listen to the coyotes calling to each other across the arroyos not far from the house. As we listen from the doorstep, my grandson whispers, "I'm a little bit afraid," but his eyes flash with more excitement than fear. Early the next morning, I see a coyote emerge from the bushes growing in the arroyo and call my mom to watch it with me. As I scan the snowy field toward the horizon, another coyote comes into view. The pair of them saunter along in the cold sunshine, seeming in no hurry to get anywhere. I wish my grandson were still here. At three, he has yet to see a real coyote, even though he sleeps with a stuffed one he named Little Howler. When he's older, he can spend the night at the farm. Surely, we'll hear coyotes yipping and yapping like a pack of rowdy teenagers strolling in the dark night.

But coyotes aren't the only animals on my grandson's mind for a farm visit. When I tell him that we can keep a lookout for coyotes on the farm when he next comes to visit, his eyes sparkle as he adds, "And bears!" I smile, but I don't make any promises. I'd love for him to see a bear at the farm, but I know it's best for all of us if he doesn't.

Our chances for survival on this planet grow slimmer every day. By "our," I mean humans, but the fate of most species doesn't look any brighter; if the planet keeps warming, we'll take them down with us, if not before us. This year's food collapse has taught me that the fate of bears and the fate of humans aren't separate. If there's no food for bears, they'll look for food where they can find it: human habitat. But no food for bears means our food supply is also endangered. If the loss continues, we'll be in peril together. Probably, we already are.

It's tragic that human actions are degrading bears' natural habitat and threatening their survival, yet in a sad way, I'm comforted that our fate is entangled with bears. Maybe paying attention to bears will help us understand the precious balance of our ecosystems, the crucial ecological kinship between threatened species, and the urgency in protecting the environments that support us all. If we don't, we'll have more bears in our backyards and more encounters to remind us to quit acting selfishly because we're not alone. In the struggle to save the future from a cage of our own defeat, it's good to know that bears are nosing around out there. ❧

Hollow | CAROLINE MILLER
Digital photography, 2015

Polistes carolina

wood pulp caked with saliva
        forms the nest, hexagon cells
translating a kind of language,
        how foundress feeds caterpillar
cud to larvae, how her triangle
        head slides into each paper
compartment the way a ball mount
        fits trailer hitch, her black
wings folded, akimbo katanas,
        not largest but most fit, though
her men permitted to procreate
        elsewhere if it means survival—
the sound of aerosol on screen
        can secure in my hand,
index finger holding down trigger.

JEFFREY FLANNERY

The Quarry | FICTION

EW IN THIS TOWN have ever seen the ocean. In fact, far easier for them to imagine the more distant and exotic plains of Africa, the dust of which is blown around the world to kindle Midwestern sunsets and litter the farmers' window-sills, than to comprehend such a thing as deeply liquid and darkly immense as the sea. The African plains, much like the Midwestern fields and prairies, offer a foothold on this earth: they bend to shovel and sweat, project a flat and steady existence that in some ways can be trusted. The ocean with its dark veneer and forbidding deep is not intuitively understood and so not trusted. Yet for billions of years, from its waters, from its depths, life viewed the world through the ocean's dense and cloudy lens. We haven't the faintest conception of what the ocean looks like down in its depths. We only know what we see, and so for the most part we don't know the sea. And this vast tiding mother-of-all returns our disinterest. So how hard is it for a person raised on these ordained plains—a person who has dug up the rich earth and dutifully planted his crops, who has looked out across endless fields that he has cultivated under his geometries, who has diverted streams, dammed rivers, and cut roads at his will and for his purpose—how hard is it for this landlocked person to believe that the earth was not, in fact, created to support human life? And so how hard is it then for him to truly comprehend the sea?

She, on the other hand, lives by the ocean. Which is to say, she lives by its rules. She has lived by the breath and pulse of the ocean ever since, as little more than a child, she realized that even here in this Midwestern town, deep in the land that is the furthest from any seashore, there is an ocean below her. Embedded in the rock of her grandfather's quarry, she discovered the thousands of jellies floating through these layers of ancient sediment. When she separated the loose layers of shale that housed them, they were not dead at all but came flooding out towards her as if freed from five hundred million years of static captivity.

They were in such numbers and in such density that she dreamed for many months of the floating black jellies falling upon her out of the darkness of her sleep.

LIKE THE SEA, the sky is hospitable only at its very edge. Some days the Midwestern sky is a calm flawless blue without depth or substance. Then there are the days when the sea above roils, fissures, and explodes with blinding flashes, releasing the load of its lakes, drowning the dusty ground in a pelting rage, punishing it with hail or covering it with snow. Braced between the barns and silos, surrounded by the faithful oak trees, the farmers ride out the fury like fisherman, each homestead an ark, a weathered and experienced vessel that has made it through these storms before.

When she had the chance, she took to the air and left the land above those dead and long-stilled oceans to study the waters on which the living seas appear. At first, through lecture and textbook, she traced the perfect explanations for waves and clouds, currents and coastlines, evolution of fins, eyes, and shell formations. But she ached to see and touch the stinging imperfections of the world, to seek out the patterns that mathematics pointed to but could not fully reveal.

BENEATH THIS SHIFTING, upturned sea mirage, the earth on which this small town was built is an old but regular patch of ground. Low-lying hills are painted on the horizon, but these rises are tired and worn, their exposed edges and rocky extremes ground down, softened by the countless arrivals and departures of the seasons. This is a place that knows too well the tiring but ceaseless march towards death. Seasons are the cycles that bring life from death, but it is not an unending cycle. With each turn from winter to spring the struggle intensifies, the miracle lessens, faith recedes.

She discovered that the most exciting objects of study, such as the great white shark or the leviathan blue whale, were also the most rare and that the researchers dedicated to these beasts spent their time waiting, and waiting. She had neither time nor patience for this. So she sought out what she could see and study in abundance. She followed the blooms and flows of the common jellies, brainless clumps of gelatin unchanged for millions of years. These beautiful soul blobs have survived ice ages, tepid seas, meteor strikes, mass extinctions, and now, man. They have shut down nuclear power stations, disabled our most powerful submarines, rendered entire nations dark and powerless. While other species have given it their all learning to fly and breathe, evolving brains and opposable thumbs, jellies just continue on as they are.

THE ANCESTORS OF THOSE who populate this town did not come here for its brutal but beautiful isolation. No, they did not settle here for its hills flush with deer and pheasants, nor for its broad flowing rivers with fish and waterfowl. Instead they came for something black and filthy that stains you down to your inner rings. Over time, continents shifted, the land rose, and the seas receded to their distant borders. And the people, they came for what was exposed in outcroppings and river banks, the stuff that lay at various depths beneath the soil, the tarsoft coal with which they could fire their smelters and mills, distill down into coke, from which they could fashion an industry of iron and steel.

Of all the jellies, *Turritopsis nutricula* fascinates her like no other. The jellyfish is immortal. That is, it constantly rejuvenates itself, transforming from a juvenile into an adult and then reversing that process back to a less developed stage. Though found throughout the world, these jellies are being studied for the most part by a few unfunded scientists in a small laboratory off a sharp-cliffed Japanese island. Two hundred feet beneath the sign that warns would-be jumpers—*Stop! A dead flower will never bloom again!*—scientists are engaged in a quest for a clue to endless life, a sea flower that dies and blooms again. It is not hard to dream that in this small jelly, no larger than the nail on her pinky, is the genetic key to human life everafter.

OPPORTUNITY. THIS IS what brought these families here two and a half centuries ago. Opportuni-

ty was written into the contracts that people signed and paid to cross the Atlantic, it was bartered at the wharves for cows and children, it was shouted by the itinerant preachers who offered pages from a book to be planted like seeds in the soils of the New World. It was bred into later generations as what they should seek, what they must find; it was whispered in their private supplications to their god; it was spoken to as if a guest at the dinner table; it was celebrated at dances, cursed and chastised at bars. Women challenged it as they stood knee deep in mud as they planted, men praised it when they stepped out onto a porch on that first spring day; it was the low last utterance, a parent's joy.

She spent enough time with the men and women who live on the ocean, the sailors, the fishermen, the men and women who toil here, to learn that the sea is not the ocean, but is the embodiment of the spirit of the ocean. The sea is its mood, its demeanor, and at times the materialization of its force. Seas appear and vanish, come and go, take lives and sometimes give them back. The ocean is the face on which seas find expression.

SIMPLE PERHAPS in lifestyle and manner, the people of this town are hardly simple to understand. Bred like Thoroughbreds on pain and disappointment, but fed fat as Berkshires on hope and salvation, the contradictions in these people lie deep. This is a people hodgepodged of heritage, language, religion, and custom, yet who meet strangers with burning suspicion, even ready violence. These people will stake their very lives for their individual rights yet have little or no tolerance for those who wish to follow their own individual beat. They are thrifty to a fault: they will smoke and cure their own meats, can and jar their summer vegetables, mend their old socks, have but one suit hanging in the closet good for weddings, graduations, funerals, and Christmas dinner, buy everything in packs of twelve and twenty-four, save the meat and vegetable scraps to feed the dogs and stray cats outside, hoard old engine oil to spray down the dirt driveway in the summer, hang their clothes in the wind instead of using electric dryers, turn off their tractor engines while driving down a hill to save a few cents of diesel. Yet they are bowed by a life-choking yoke of debt on farmland, crops, machinery, big screen TVs, ATVs, and river boats. And out of

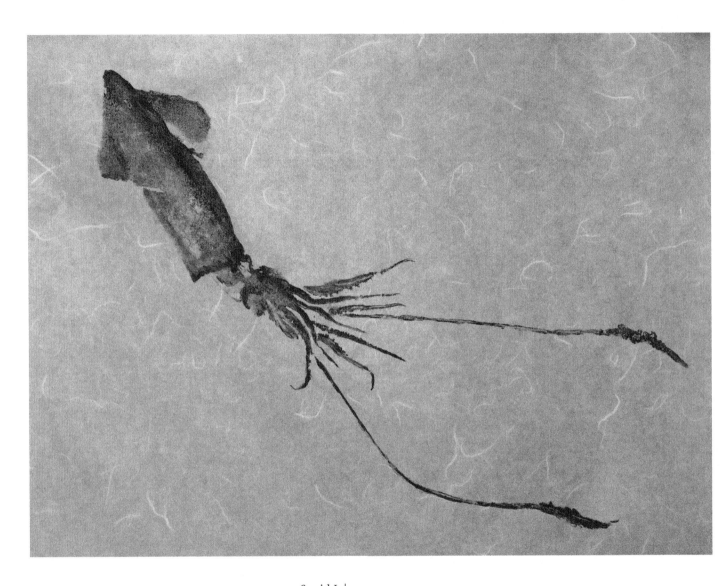

Squid I | ALYSSA IRIZARRY
Squid and block printing ink on paper, 18 x 24 in., 2016

the same spirit they condemn the lazy and the idle, they never complain of the toll work takes on life and the living: the missing fingers, broken ankles, bowed backs, or stone-hard lungs.

She learned that there are lessons in the earth, and that the farther you go back—back to what the scientists call the *dark time* when animals were only cell membranes wrapped around water—the stronger those lessons are. A story she considers biblical in its impact took place some three billion years ago when the cyanobacteria rose up and conquered the earth. Suddenly, life assumed the ability to create energy from the sun. Life could finally creep forth from the depths, the hot fissures, the volcanic nooks far beneath the surface. The cyanobacteria consumed carbon dioxide and excreted oxygen. It oxidized metals—rusted the earth. It joined with other organisms to become the chloroplasts in plants and the eukaryotic algae. These were the mounds of ancient civilizations that ruled the earth for billions of years.

BY MOST ACCOUNTS, Midwesterners are a polite and accommodating people, their voices low and measured, greetings pleasant and answers considerate. But their demeanors shade a core that has been repeatedly broken and repaired, split and healed, each time creating a scarred caution in attitude and belief that can only be submerged so long. And so, in the early hours of the morning, those inner voices rise from the dark oceanwells of minds, and the radio crackles with the rants and condemnations of the silent witnesses who come forth to warn and condemn those who ignore the crop signs and abductions by aliens, who allow the corruption and waste of our governments, and who cannot see the long, shadowy reach of the devil's claw into every aspect of our lives.

Her scars: the deep purple slug across her kneecap from when she fell against a skeleton of coral, the delicate loop of red seaweed around the base of a thumb she got caught in a sailboat's jibbing, the graygreen seahorse tail that grasped her chin when a wave threw her face first into a pillar of rock, the white starfish on her shoulders from the day she was too careless in a Mediterranean sun, the milky krilldrift across her corneas from all the days of squinting across the sunlit sea.

AFTER TWO HUNDRED YEARS of ceaseless digging, nothing remains of those layers of ancient fecundity but dry veins and piled dregs. The fiery smoke from the mills is gone, pubs no longer clog every street corner, the red light districts were shut down and cleared away, the revival tents have stopped visiting the baptismal waters of the river banks. The itinerant medical healers, the recruiters for the railroad, the men with pomade-slick hair selling options on land in California are all historical oddities now. The Victorian mansions along tree-lined avenues are no longer occupied by the elite; if not turned into museums or nursing schools, most have been shuttered, too expensive to heat. The houses built by the factories for the working mugs still sit like upturned stones on the low hills and asphalt roads weave their way across and back and forth. Churches and banks are still blackened by soot and most sit empty now. Lifeless factories tattooed with graffiti crouch darkly off the river where barges once floated, and warehouses sit hollow between the ancient iron bridges that once rattled and creaked with the weight of trains carrying their loads of coal and ore.

She had come to realize that we are mere accidents upon this planet. We are at the mercy of such small things: a degree or two in temperature, a point or two rise in acidity. The movement of a current. Any difference in what had been and we would not be here. Yes, we are even smaller than we think. We are minuscule in comparison to something that keeps us from destroying ourselves. It is not the trees. Not the millions of square miles of ocean. Not the vessel of air above us. And it is certainly not some grand design. Or was it? No, it is all part of the larger cycle, she would say.

AS ITS INDUSTRY DIED, the men and women here grew empty and passive while other forms of life found opportunity for renewal. The violent wounds of old mining operations and the tie-sutured gashes left by abandoned railroads began to heal, were slowly absorbed into forgiving hills and valleys. The sky lost its perpetual haze and no longer burst into flame at sunrise or sunset; it smolders now on distant berms. The streams expunged their rusty sediments, waters flow clear again over mossy rocks, and birds nest in strands of tall grass and velvet-tipped saplings.

Nature, though slow and gentle in its recovery, ignores man's ignobility.

She never imagined that one day she would return. She had always thought that her hometown was a place best forgotten. In some ways she felt that she had traveled the world and all she had seen and felt, all that had hurt and pleased her, was now inside her. She returned home with enough money to buy an old building in the industrial part of town, a building she converted into a space that is empty. She bides her time in this town. Her travels now take place in her brain, with her eyes closed, during sleep. She prefers to remain in place, to let the world revolve around her for a change while she plays the role of point of reference.

TODAY, NEW METHODS troll the previously forbidden deposits in the black shale. The drills sink not just deeper but snake horizontally, crisscrossing the strata beneath the farmlands in search of the gaseous stone. A new prosperity has returned. New wealth comes to those who were dumb enough to hold onto their once worthless land. And once again flares burn and lighted rigs flicker from fields and farms. Work sites are cleared overnight on the farms that have signed the leases with the oil companies. Machinery is brought in and installed within hours; drilling begins within days. Shiny stainless steel tankers roll nonstop, transporting water, some say one hundred million gallons a day, and top loaders carry millions of pounds of sand to be pumped deep into the rock until it cracks and spits up.

During the day she picks through the dust and broken flagstone of the quarry, looking for more of the captured *medusae*, the two- to three-foot diameter jellies that were stranded so quickly on some sandy beach that their last breaths and movements are recorded in rock. These fossils begin to fill her home. All facing the same direction, they form nearly perfect rings in the flagstone.

THE VOICES AT NIGHT offer a confused banter these days. The cries decrying Satan's deeds have slipped beneath the hum of the universe, and conspiracies have crawled back to whence they came. Progress has touched them all and awakened a new faith in capital. No devil seems significant in light of opportunity.

There is always a trade with the devil, she now believes. We humans gave up a long life on earth for brains that gave us a quick chance at understanding everything. She thinks we overestimated our bet. The jellies, on the other hand, traded simplicity for immortality. And yet as much as she desires her own immortality, she recognizes the world will never follow suit. And just like humans, the immortal jellyfish will do what they were designed to do: They will feed and they will bloom, they will drift like a living snow and gently and graciously suffocate the world that, on a day no one will count, will reawaken and carry on. ❧

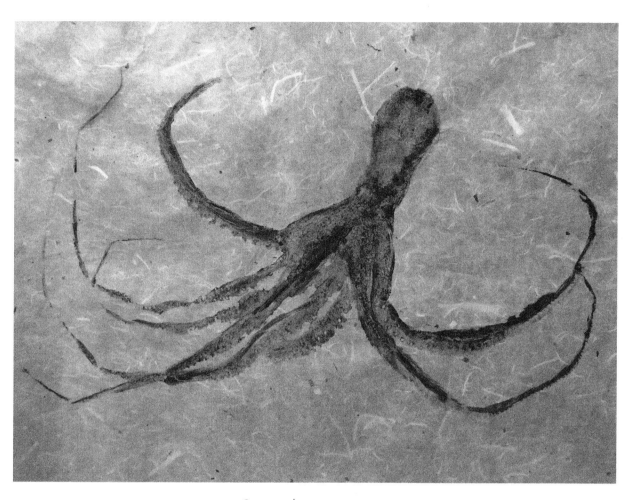

Octopus | ALYSSA IRIZARRY
Octopus and block printing ink on paper, 18 x 24 in., 2016

Badlands | CAROLINE MILLER
Digital photography, 2015

contrails

be scientific, be intellectual

lecture me on cirrus aviaticus
how hot exhaust and particulates
condense and dance and disperse
over seconds, minutes, days, weeks—
tickle my brain with explanations
of contrail climate change

be magical, be mysterious

whisper to me how the ghosts
of every airline disaster
float behind those jets
billowy white fingers reaching
stretching for miles
clawing their way aboard

be ignorant, be earthbound

explain to me how you never once
look up to see those water lines
raking fingernails across a chalkboard sky
never bother with things so far away
when each spring's preoccupation
is that the creek will rise too high
flooding out the garden sprouts
each summer is spent fretting
that it will run too low
and all the corn will wither

be lyrical, be poetic

write lines that catch my breath
before i exhale into winter air
my own vapor trail that marks
if only for an instant
my passage through your life

BLENDING

We mixed the musts of our sweet and bitter apples. You brought the blend to your mouth and then to mine. In it, I could name the colors of familial love, the taste of nostalgia, the warmth of pleasure. I asked what you tasted, and you said, home.

Cuttings | NONFICTION

I F THERE WAS EVER A PINECONE, my granddaddy would find it. He was up at dawn every morning, a nubby navy blue bathrobe rough against his skin, the two fingers and thumb that remained on his right hand wrapped around the handle of a coffee mug. He would slide the glass door to one side, step out on the small brick patio with a newspaper under one arm, close the door firmly, settle into the black iron chair.

Later, the iron would be hot to the touch—hot enough to burn red criss-cross marks on the backs of a careless granddaughter's thighs. But now, a ritual meant for the cool morning: sip fresh coffee, page through the paper. Survey the land.

After the coffee and before breakfast was the rough groan of the tractor, its passing growl a warning that I had slept in. For a few minutes I might be able to watch the ceiling fan, running my hands over the fringed bedspread. Then the sharp crack of my grandmother's voice—immediate reason to scramble down the four cold brick steps to the open den, up the two steps to the kitchen, and arrange silver on the grass tablemats before the bells on the door clanked and granddaddy came in stomping, smells of oil and dog food following him from the garage. And after that I could run barefoot through the lawn, nary a pinecone to bruise my toes. Careful only of scattered fire ant nests and the dogs.

Not these dogs, though. Not yippy, fuzzy mutts. I cough, wrench my arms from the dirt, roll over, shimmy out from under a mass of vines. Sweat has drained directly onto the insides of my glasses, and everything I might wipe them with is smeared with clay. I shake them instead, the fence and the dogs and the weeds temporarily blurred. I squint, and for a moment the lawn looks as even as I remember. But brown.

The dogs keep barking.

"You," I sigh tiredly, "are not scary. Do you have any idea who used to live here?" They probably don't;

smells only linger so long. "Rush could eat you in one bite. Hell, he could eat me in less than five."

Someone always called me in before the dogs came out. Because of Rush, they said. It confused me—why would Rush hurt me? I remember my dad's laughter. "Because he's been trained to. You start running, and that dog will be on you like *this*." His fingers enormous in my face, the snap resonant.

A fierce dog for my gentle grandfather, but the Rottweiler was here because people kept stealing the Labs. Southern logic.

I squint through my smeared lenses toward the driveway. No sign of a car yet, thank God. I'm trespassing, and I know just enough about the bad blood surrounding the sale to know that citing my lineage is not going to earn me any privileges. Especially given that I'm clotted with dirt from head to ankle, digging up a past that, by rights, does not belong to me.

The dogs keep barking. I sigh again, rock back on my heels. Dry, wiry grass stabs my exposed calves.

I survey the land.

Mother had been wistful, her gloved hands resting on the top of her garden shovel. I'd looked up when the scrape of metal on clay had suddenly stopped, found her staring out over the plot we'd halfway cleared. Her face was turned away, but I knew her eyes were filling out the graph paper on her desk: its imagined flowers and fruits, its layered seasons and their colors.

"I wonder," she mused softly, her eyes somewhere between the blueberry bushes and the buckeyes, "if muscadine wouldn't grow here?"

It was a silly question; we'd been uprooting the wild grapevines all week, saving their space for the squash and tomatoes. In the pause, I became conscious of the humidity, the heat building despite the calendar month. We would have to go in soon.

"Muscadine? Mom, it's all over the place."

"It can be finicky. Unless it's already established."

"Oh."

She half turned toward me, frowning at the far side of the plot. I remembered that I would be driving past Matthewsville the next day. Past granddaddy's old homestead. I looked at my mother, at the wrinkles around her eyes, at her heavy, drooping gloves.

This is her last garden. The one she won't leave behind.

I straightened, yawned, thought for a moment. Then I shrugged.

"She moved the fence in, right?"

They say a Southern lady can marry only four types: a doctor, a lawyer, a preacher, or a military officer. My mother got two out of four. The military, the ministry, my dad's itchy feet—every couple of years, one or another uprooted our family.

My mother is not a quiet woman, but she is an agreeable one; she is not a pliable woman, but she is a supportive one. A typical middle child. Still, maybe there was something of a rebellion in our pilgrimages to the Carolinas each year, taken despite expense and my dad's churlishness. And maybe there was some rebellion in the cuttings and seedlings that blocked our rear windows on each drive back, making my father grind his teeth and swear whenever we hit traffic.

Something from granddaddy's garden always found its way into our backyard, no matter how improbable the climate zone. The azaleas took just fine in Louisiana, of course, and the pine trees sprung up fast as weeds in Spartanburg; but there were also the hydrangeas under the waterspout in Illinois, the hostas ringing the Kansas porch. Moving meant choosing colors—first the colors of our walls, and then the colors in the garden. Early on I came to understand the importance of balance: tall plants in back and short ones in front, bright plants in shadows, the graph paper swarming with what blooms when and where. All year long we should have color, should have depth.

By the time I hit age seventeen, the process had become less exciting and more routine. The gardens were planted, but less regularly tended; the walls were painted a little less brightly, with a little less care. But there was always the sense of possibility in talking over a plot of earth, always my mother's hands spreading, swooping, framing pieces of the yard, her forehead slightly bunched, her eyes alight.

From my father, I got itchy feet and a habit of cursing in traffic.

My mother gave me her vision.

I try to use that vision now, but it's no good. There are four-foot-high weeds around the buried septic system—whether thanks to water or shit, I have no idea. Elsewhere, the weeds are shorter but no less thick. Everything brown, and all but two of the trees gone. I remember mother's quiet crying as we walked the lawn, pink ribbons in our hands, trying to find some worth saving. "Something has to be left." Her eyes straight ahead, not scanning, barely blinking. "Something has to be left."

That was after the ice storm snapped the full-grown pines—forty of them, the ones my grandfather had planted when he and grandmother got married—like toothpicks. It was before grandmother called to let us know, with her customary steel, that she was giving up on the farm. Moving into the mountains, away from the heat and the family feuds. Away from the enormous, graying lawn, that—even after she'd moved the fence in—no one could properly care for.

The dogs are still barking.

I remember running, running, running, running through the grass that was always even and so endless. I remember spinning until the dark pines and the blue sky blurred. I remember collapsing into a tilting world, fingers laced through the grass, the trees creaking over the sound of my own breath and laughter.

I speak loudly. "I have learned that lawns are artificial. Homogenous. Unhealthy." A gold ring glints through the dirt on my right hand, the black stone in it so like my mother's. "They're *monocultures*."

The dogs keep barking.

I turn back to my digging.

Here's another thing I remember: sitting in my Sunday dress, getting dirt over my black patent-leather shoes and white lacy socks, and it wasn't clay dirt but black dirt. Heart-shaped saw-edged leaves dense over my head. A blurred buzzing of wasps. I rocked from side to side checking for sky, but the leaves only gave me shades of green.

We Were Here: Cigarette/Deadheads │ NATALIE VESTIN
Ink on paper, 4 1/2 x 6 in., 2016

We Were Here: Snake/Mandible | NATALIE VESTIN
Ink on paper, 4 1/2 x 6 in., 2016

Like as not, I was singing.

Outside, calls: to lunch, to play. Any time I could crawl out from under the vines and run to meet them. Any time I wanted.

Instead, I reached my hand slowly through the wasps, looking for the fruit. Not for eating, yet; not until later, when they turned dark. It was granddaddy who taught me how to bite through the thick, leathery skin—just a piece of it—and suck the tangy sweetness from inside. Muscadine: a name suited to a Southern drawl.

"It doesn't taste like anything else," I'd said, and he'd smiled his easy smile.

"Right you are, shugga. Right you are."

Not for eating. Not yet. But my hand reached out, slowly, again and again, looking for the fruit.

I was only nervous for the first half hour or so. Then it became easier to focus on my fingers groping blindly for the rough curve of the root, the earth swelling tight and moist against my forearms. I occasionally wiped my forehead against my hunched shoulders, grinding dirt into the sticky sweat around my temples; I occasionally wiggled my torso deeper into the dirt, trying to scatter the critters crawling over my legs and back.

This does not, by rights, belong to me—a girl with itchy feet, who knows most about plants when they are colors on paper, who struggles to get beyond vision. A girl who, faced with two twining spires of vine each the size of her calves, thoughtlessly assumed a taproot. I'd cut around and then through the first, coming away with a rootless stick, growth that took four of my lifetimes completely severed. I'd cried, briefly and furiously, the beautiful canopy of my childhood browning and crumbling in my memory. "I'm so sorry. I didn't mean to. I'm so sorry."

Those damn dogs barking at my back.

Now I am focused, trying to draw out this thing that is so much older than I am. Coaxing it into a waiting burlap bag that I will carry through the pine woods and down the old lumber trail. I'll sweat and curse and heave it into my car, try unsuccessfully to brush the dirt from my clothes. Close the door. Drive with the windows down, squinting through my glasses and singing.

All the way to my mother's garden, where she works between coffee and breakfast. ❧

Ghazal: Quilt

Here hangs the horizon-long line with which you tied the first song
to a tree like a tire swing for me to find and fly on—

There beyond the busy bank where we dip a bowl of grapes into the cold
runs the current-slap like satin slippers we can try on—

Here among eight hollows holding prayer-hewn history close
a hallowed cave holds out a boulder built to cry on—

There between potato-fields greening summer your laugh and my tipped pen
sip the sun-spelled libations they've been high on—

Here Picture Rocks bottle unfiltered paints, shake, uncork, and spread
ecstatic bed of sky fit for a poem to fall and die on—

There your voice and my disordered lines dive into pool of saffron tea,
deep in search of rare choral reef of words you can rely on—

Here, wisps of jellyfish clouds kiss the cliff's outstretched pinkie and we,
content with a tiny square of quilt for us to lie on.

Living in the Edge | NONFICTION

THE FIRST SUMMER MY HUSBAND AND I were together, we lived for a number of weeks in a borrowed tipi. My friend Debi had loaned the tipi to us with great excitement. A former lover of hers had lived in the structure most recently, and she was happy to arrange transporting the rolled canvas and long poles. The morning it arrived at the field we had also borrowed, I spiked a fever. I lay in the opened hatch of an old Saab we owned then, and E. checked in on me periodically. I vaguely recall the positioning of the poles was quite precise, and unrolling the heavy canvas must have been difficult. I had a fever of intensity where I was no longer wholly in my body, but floating elsewhere, part of the current scene yet simultaneously not part of it. The whole unfolding of this house seemed to have happened without my participation.

Living in a tipi was like nothing else for a number of reasons, the most immediate being (and this should have been a giant *well, duh*) the base of a tipi is *round*. Risking sounding like the freakiest hippie, I'd describe this difference as a profound change in energy flow. That tipi was quite large, with a soaringly high center. Lying on our backs on the bed, our entire domestic life was then contained within that circle, arched over by the poles lodged in the earth and joined in a knotted clasp over our heads. Energy ran around the circle, rather than bollixing in corners. A week into tipi sleeping illuminated an experience of the world I had never imagined; with the dissolution of walls, the world appeared surprisingly unfettered. Tipis, of course, with their buffalo origins, were unknown to this area's indigenous Mahican and Pennacook, who lived in longhouses. In a post-modernist nomadic style, this tipi was a contemporary version of beige canvas, but, nonetheless, something that could be stashed in the rear hatch of a decrepit Saab.

Living in a summer-world tipi is living in a realm of edge habitat. While more permanent tipi dwelling—and certainly tipi life in colder temperatures—would require sealing the tipi skirt to the earth, we simply laid the canvas on that blueberry field and spread a carpet remnant in the kitchen area. On sunny days, we rolled up the sides and let the unchecked breeze flow through. Rainy days, we listened to pattering on the rolled-down canvas, the only portal to the world the small domed door whose flap we frequently propped open. Edge is the joining of two places—field and stream, forest and meadow—that mingling of diversity where wildlife thrives, where songbirds nest, groundhogs tunnel, foxes hunt. Where the heavy drape of canvas lay on the ground became a variant of edge habitat for us. Roll it up, and we could bend out and pick blueberries for breakfast oatmeal. After sunset, the far side of the tipi, where we stored odd things, was filled with darkness and night, while we cooked at the plywood board of a kitchen counter or lay in bed with the glowing smallness of hurricane lamps. One night, a visiting skunk brushed E.'s bare knee as he lay in bed, and we jumped out in the dark and drove down the field, where we slept the night's remainder in the back of our Rabbit. The next day, we sprinkled moth balls around the tipi to keep this marauder away.

Thinking back now, with a repugnance of naphthalene, I wonder at those days, when the edges of our world seemed to be filled with nothing but sunlight and cicadas, my fear of the dark, and harmless—albeit sometimes odiferous—wildlife. In those days, with great drama, we believed what plagued our sleep might be remedied with a box of white flakes we purchased in the grocery store.

A singular element of that energy flow in my tipi-dwelling days, no doubt, was the starriness of untried love. We were at the nascent place in our relationship where we believed our future would always sing with joy. In those Vermont summer months, I savored the lengthy sun-lit days, the tipi sides tied up so the fragrant season moved right in.

It's now been so many years since those days. I look back on my life as a young woman and see myself as not much more than clay just beginning to breathe and push. Certainly, the love E. and I had was tender, like a garden sprouting up through stony earth. A friend's little girl, maybe three years old, with straw-blond hair and uneven bangs, wandered through our group in those days. Her parents were split up. The father's new girlfriend birthed a son that July. The little girl carried around a wooden Ben & Jerry's truck and held it against her skinny chest, nursing her toy, one of my keenest memories of that summer. How much I desired a baby, a child to solidify E. and me from *couple* to *family*.

Since then, E. and I have lived and slept in so many different spaces, all over this continent, from a soulless metal trailer in Washington State to the mouse-inhabited hunting camp we bought years later in wooded northern Vermont. Within me, I carry all those structures. One snowy Halloween, camping in Yellowstone, a ranger woke us at dawn, knocking on the window of the black diesel Rabbit where we slept. We believed we were snagged for slinking in without paying, but instead he politely handed in our little cooler, which held nothing but some foully fermenting chickpeas, and cautioned us about bears. Hadn't we read the signs posted everywhere? Indeed, we had not. Even now, we laugh about that, how kind the ranger was—not at all what we expected—and how cold that mountain air was on our skin.

Our life has sprawled so far from those saccharine days. We now have two beloved daughters, birthed by the scalpel. We've gone through a vale of tears and recrimination and misery; we've weathered through too-long days, through nights of illness, through injury and uncertainty and stark fear. So deeply into this marriage now, when I think back on those days that seemed so lusciously eternal, I remember the scent of that dirt path winding through the back part of that field. I had never lived beyond the pavement before, and the dust rose in those hot summer days, the earth giving itself up to the sky. The blueberry bushes were too scrubby and wild to produce more than meager handfuls of berries, and when it rained, I staggered awake in the middle of the night and stashed the library books in the car, as water inevitably dripped down the poles. At night, I read by two kerosene lamps placed side by side, lousy for reading but unmatched in warmth. I learned to clean the glass globes every morning with crumpled newspaper. Since those days, our energy has repeatedly rattled unhappily in corners, squirreled around, not at all creative, not at all imbued with that holy transcendence of those wooden poles, ends on the ground, tips crossed in the heavens. Our life has been as profane to the core as possible.

And yet, I resist the over-idealization of that round house, of the blush of early love. Our marriage is by far a grittier thing; as clouds hold hailstorms, birth demands blood, and every seed contains its own life and death. In my youthful days, I saw those discrete things—tipi, field, glass lamp—as things unto themselves. But human life, intensified through relationships, is all edge—hunt and home, blueberry and skunk—dynamic, alive, shifting. Tipis are designed to be rolled up and carried from place to place, a movable hearth, a home between our hands, a domesticity and family life continuously rekindled and re-imagined. A life inherently always in the re-making. ❧

Hooded View of Death Valley Crater | WILLIAM C. CRAWFORD
Digital photography, 2015

Fruit trees next door

These persimmons have no more to say
about green. From a distance,
you can catch a barely audible yellow,
though not as tight-lipped as lemon.

But up close, wetting a thumb
to wipe the dust off,
you can hear the ochre in it.

And some outgoing fruit
has begun to whisper
about a reddish-brown orange—
only a rumor so far
that the whole tree will confirm
come November.

Palm Dairy

The light is almost liquid
this early, infused
with that indigenous herb
we call Stillness.

A cool tea to wake up with,
it has the aroma of
alfalfa cut late last night,
a slightly saffron color,

and a taste
I don't know how to describe—
but want to sip again
tomorrow.

CONTRIBUTORS

Maggie Blake Bailey has published poems in *The Southern Poetry Anthology, Volume V: Georgia*; *Tar River*; *Slipstream*; and elsewhere. She has been nominated twice for a Pushcart Prize and her chapbook, *Bury the Lede*, is available from Finishing Line Press. Her website is maggieblakebailey.com.

Brian D. Cohen's artist's books and prints have been shown in forty individual exhibitions, including a retrospective in 1997 at the Fresno Art Museum, and he has participated in over 150 group shows. Cohen's books and etchings are held by major private and public collections throughout the country, including The New York Public Library, the Library of Congress, and the Philadelphia and Portland (Oregon) Museums of Art, as well as the United States Ambassador's residence in Egypt. He was the first-place winner of major international print competitions in San Diego, Philadelphia, and Washington, DC.

William C. Crawford is a writer and photographer who lives in Winston-Salem, NC. He was a combat photojournalist in Vietnam. He later enjoyed a long career in social work and taught at UNC Chapel Hill. He photographs the trite, the trivial, and the mundane. Crawford developed the forensic foraging technique of photography with his colleague, Sydney lens-man Jim Provencher.

Jeffrey Flannery divides his time between Natchez, MS, and Minneapolis, MN. Aside from writing about environmental issues and ideas, Jeffrey has worked for years developing new technologies that support a more sustainable world. His short stories have appeared in *34thParallel*, *Five2One*, *The Dark Mountain Project*, *Semaphore*, *Ducts Journal*, *Deepwater Literary Journal*, *Electric Windmill Press*, and other literary outlets. Passionate about storytelling, Jeffrey often participates at The Moth and other StorySLAM events.

Benjamin Goodridge is a freelance writer living in Portland, ME, with his fiancé, Jackie. He has published multiple articles and enjoys writing fiction with every free moment he has.

Alyssa Irizarry studied environmental studies, art history, and visual arts at Tufts University. She is the program manager at Bow Seat Ocean Awareness Programs, a nonprofit that promotes ocean conservation and advocacy through the arts. Her research on environmental muralism is featured in an exhibit at the Monterey Bay Aquarium in California. She lives in Salem, MA.

Talley V. Kayser is an outdoor educator, literature teacher, renegade scholar, and rock climber. She recently returned to the South following several years out West. She holds an MA in literature and the environment from the University of Nevada, Reno.

Jenny Kendler is an interdisciplinary artist, environmental activist, naturalist, wild forager, and social entrepreneur. She is the first artist-in-residence with environmental nonprofit NRDC. Her work has been exhibited nationally and internationally at museums and public venues. She is vice president of the artist residency ACRE and co-founded the artist website service OtherPeoplesPixels and The Endangered Species Print Project, which has raised over $14,000 for conservation. See more of her work at jennykendler.com.

Alexis Lathem is an environmental journalist, editor, writing instructor, and author of the poetry collection *Alphabet of Bones* (Wind Ridge, 2015). Her poems have appeared in *Hunger Mountain*, *Chelsea*, *Spoon River*, *Saranac Review*, *Beloit*, and other journals. She lives on a small farm in Vermont.

Jim Lewis is an internationally published poet, musician, and nurse practitioner. His poetry and music reflect the complexity of human interactions, sometimes drawing inspiration from his experience in healthcare. When he is not otherwise occupied, he is often on a kayak, exploring and photographing the waterways near his home in California.

Farzana Marie grew up in Chile, California, and Kazakhstan. Farzana's poetry and translations have appeared in print and online journals including *The Rusty Nail*, *Adanna*, *Fourteen Hills*, *Zócalo*, *Antiphon*, *Guernica*, *The Atticus Review*, and *The Fourth River*. She is the author of the nonfiction book *Hearts for Sale! A Buyer's Guide to Winning in Afghanistan* (Worldwide Writings, 2013), a poetry chapbook, *Letters to War and Lethe* (Finishing Line Press, 2014), and a book of Persian Dari poetry in translation, *Load Poems Like Guns: Women's Poetry from Herat, Afghanistan* (Holy Cow! Press, 2015). Find her online at farzanamarie.com and on Twitter @farzanamarie.

Sam Mass is an educator, artist, and butcher living in Nashville, TN. She is interested in the intersection of arts, social justice, and the environment, with a particular interest in sustainable food systems and food justice. She has an MA in food studies from Chatham University, where she studied urban farms and the communities that surround them.

Grant McClure is a writer from Charleston, SC. He studies English and environmental science at Wofford College, and his most recent work has appeared in *The Adroit Journal* and *The Best Teen Writing of 2015*. He received the Student Fellowship for Fiction from the SC Academy of Authors in April of 2016.

Caroline Miller is an amateur photographer living in picturesque southwest Colorado. She studied photography at Skidmore College in Saratoga Springs, NY, before moving west to work with horses and take in the landscape.

Kristen M. Ploetz is a writer and former land use attorney living in Massachusetts. Her other writing has appeared in print and online journals including *NYT Motherlode*, *Literary Mama*, *Modern Farmer*, *The Humanist*, and *Manifest-Station*, among others. She is currently working on a collection of short stories. Find her online at littlelodestar.com and on Twitter @littlelodestar.

Erin Elkins Radcliffe's poems have recently appeared or are forthcoming in *Tupelo Quarterly*, *San Pedro River Review*, *Rogue Agent*, and *Coal Hill Review*. Originally from Indiana, Erin lives in Albuquerque, NM, with her family. More of Erin's work can be found at erinelkinsradcliffe.com.

Meghan Rigali graduated from the San Francisco Art Institute in 2003 with a BFA in interdisciplinary studies, combining classical and conceptual lineages in visual art. Work as a wilderness therapy guide with at-risk youth in Vermont radicalized Rigali's relationship to the natural world, a shift that continues to extend into her work as an artist, educator, and steward of Nature. Her commitment to practices in yoga, eco-depth psychology, and contemporary wilderness rites enriches her work as a middle and high school teacher of visual art in Southern Vermont. Find her online at meghanrigali.com.

Jenny Ruth's fiction has appeared in *Streetlight Magazine*. She is a freelance writer for local magazines, 5K & SoleMates Director for Girls on the Run of the Greater Susquehanna Valley, and founder of a community-based literary project called *The 522 Review*. You can find her online at jennyruthwrites.com and on Twitter @jennyruth81.

Kayann Short, PhD, is the author of *A Bushel's Worth: An Ecobiography*, a memoir of reunion with her family's farming past and a call to action for agricultural preservation today. Her work has appeared in various literary magazines, including *Pilgrimage*, *The Courier*, and *Mad River Review*, and her essay "Soil vs Dirt: A Reverie on Getting Down to Earth" appears in the recent anthology *Dirt: A Love Story*. A former faculty member at the University of Colorado Boulder, she farms, teaches, and writes at Stonebridge Farm on Colorado's Front Range.

Stephen Siperstein lives in Eugene, OR, and is a PhD candidate in the Department of English at the University of Oregon. His poetry has appeared most recently in *ISLE*, *Poecology*, and *Saltfront*, and he is the co-editor of the forthcoming volume *Teaching Climate Change in the Humanities* (Routledge, 2016). When not writing, he is usually hiking in the Cascades or tending to his backyard garden and three chickens.

J.D. Smith's fourth poetry collection, *The Killing Tree*, will be published by Finishing Line Press in July of 2016. In 2007 he was awarded a Fellowship in Poetry from the United States National Endowment for the Arts. His individual poems have appeared in publications including *Dark Mountain* and *Terrain*.

Brett Ann Stanciu is a writer, sugarmaker, and believer in home gardens and using clotheslines. She lives in northern Vermont with her two daughters. Her novel, *Hidden View*, was published by Green Writers Press in 2015. Read her blog at stonysoilvermont.com.

Kelsey Swintek is a serial dater and pizza enthusiast currently living in NYC. She graduated from the University of St Andrews with a degree in comparative literature and art history and subsequently spent time hanging bed sheets in the south of Spain. Her portfolio is influenced by Susan Sontag's *On Photography* and the aesthetics of trees.

Don Thompson was born and raised in Bakersfield, CA, and has lived in the southern San Joaquin Valley for most of his life. Now retired from teaching in the prison system, he lives with his wife, Chris, on her family's farm. Thompson has been publishing poetry since the early sixties, including a dozen books and chapbooks. For more information and links to his publications, visit his website, *San Joaquin Ink*, at don-e-thompson.com.

Natalie Vestin is a science writer and artist from Saint Paul, MN. Her essays have appeared in *The Normal School*, *The Iowa Review*, *Puerto del Sol*, and elsewhere. Her nonfiction chapbook, *Shine a light, the light won't pass*, was published in 2015 by MIEL, and her fiction and photography chapbook, *Gomorrah, Baby*, is forthcoming from Anchor & Plume in 2016.

Cynthia Scott Wandler is a writer whose work is expanding beyond the hundreds of newspaper articles she has written. She participates in Story Slam and quick-story events, placing first one glorious time. Living in Morinville, Alberta, with her husband and two children, she can often be found either writing on paper or hugging the trees that make it.